65 Structure and Bonding

Solid State Chemistry

With Contributions by
M. Ardon A. Bino J. K. Burdett P. C. Schmidt

With 87 Figures and 30 Tables

Springer-Verlag
Berlin Heidelberg New York
London Paris Tokyo

ISBN 3-540-17581-4 Springer-Verlag Berlin Heidelberg New York
ISBN 0-387-17581-4 Springer-Verlag New York Berlin Heidelberg

Library of Congress Cataloging-in-Publication Data. Solid state chemistry. (Structure and bonding; 65) Bibliography: p. Includes index. 1. Solid state chemistry. I. Ardon, Michael, 1928– .II. Series. QD461.S92 vol. 65
541.2'2 s 87-9546 [QD478] [541'.0421]
ISBN 0-387-17581-4 (U.S.)

Printed in Germany

Typesetting: Mitterweger Werksatz GmbH, 6831 Plankstadt, Germany.
Printing: Druckhaus Beltz, 6944 Hemsbach, Germany.
Bookbinding: J. Schäffer GmbH & Co.KG., 6718 Grünstadt, Germany
2152/3140-543210

Table of Contents

A New Aspect of Hydrolysis of Metal Ions. The Hydrogen-Oxide Bridging Ligand ($H_3O_2^-$)
M. Ardon, A. Bino 1

Some Structural Problems Examined Using the Method of Moments
J. K. Burdett 29

Electronic Structure of Intermetallic B 32 Type Zintl Phases
P. C. Schmidt 91

Author Index Volumes 1–65 135

A New Aspect of Hydrolysis of Metal Ions: The Hydrogen-Oxide Bridging Ligand ($H_3O_2^-$)

Michael Ardon and Avi Bino

Department of Inorganic and Analytical Chemistry, The Hebrew University of Jerusalem, 91904 Jerusalem, Israel

This review presents a new structural feature related to the hydrolysis of metal ions, which has been revealed during the past four years. It concerns a new form of bridging between metal ions. An OH ligand coordinated to a metal atom may form a short and symmetrical hydrogen bond to an H_2O ligand of another metal atom, thereby forming a hydrogen oxide bridging ligand ($H_3O_2^-$) between these atoms. This ligand was first discovered and recognized as such in 1981 by X-ray diffraction studies of some crystalline salts of trinuclear triangular cluster ions of molybdenum and tungsten. It has since been shown to exist also in classical coordination compounds. This ligand plays a fundamental role in the structure of primary hydrolysis products of metal ions. Some of these products, formerly believed to be mononuclear, such as the socalled hydroxoaqua ions $[ML_4(H_2O)(OH)]^{(n-1)+}$ were shown to be dimers or chain-polymers in which the metal atoms are bridged by hydrogen oxide ligands. While X-ray studies established the existence of hydrogen oxide bridged dimers in crystalline hydrolysis products of metal ions, other physico-chemical methods were used to demonstrate the persistence of these structures in aqueous solutions. The impact of these findings on the overall picture of hydrolysis and in particular on reaction mechanisms of redox and substitution reactions is discussed.

I. Introduction . . . 2

II. Very Strong O–H–O Bonds . . . 3

III. $H_3O_2^-$ Bridging in Trinuclear Clusters . . . 5

IV. Hydroxoaqua Metal Ions . . . 12

V. Aqua-Hydroxo Double Salts . . . 18

VI. Hydroxoaqua Metal Ions in Solution . . . 20

VII. Mechanism of Reactions Involving $H_3O_2^-$ Bridging . . . 22

VII. 1. Mechanism of Olation Reactions . . . 22

VII. 2. Redox Reactions . . . 25

VIII. Conclusions and Perspectives . . . 25

IX. Note Added in Proof . . . 26

X. References . . . 26

I. Introduction

The existence of a *hydrogen-oxide* bridging ligand, $H_3O_2^-$, between metal atoms was first reported in compounds of trinuclear, triangular clusters of molybdenum and tungsten[1–3]. It has since been shown to play an important and quite general role in classical coordination chemistry[4–7]. This ligand (Fig. 1) is formed by means of a short and very strong

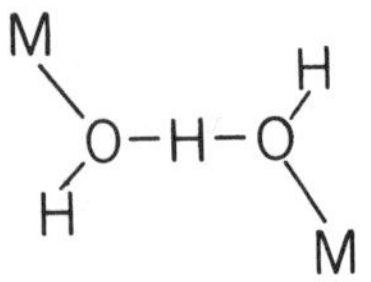

Fig. 1. The $H_3O_2^-$ bridging ligand

hydrogen bond between a *hydroxo*-ligand of one metal atom and an *aqua* ligand of another metal atom, as postulated by G. Schwarzenbach[8]. The bond length of this H-bond is, in most cases, less than 2.47 Å and the bond energy is approximately 100 kJ/mole[9, 10]. These findings call for some modification of the structural model of *hydrolysis* of metal ions, developed at the beginning of this century[11]. This classical model may be summarized as follows:

1. A metal ion, with a coordination number N, having one or more water ligands, may lose a proton of one of these ligands

   $$[ML_{N-m}(H_2O)_m]^{n+} \rightarrow [ML_{N-m}(H_2O)_{m-1}(OH)]^{(n-1)+} + H^+ \quad (1)$$

   If $m > 1$, the resulting *primary* product of hydrolysis is a *hydroxoaqua* ion such as $[Co(NH_3)_4(H_2O)(OH)]^{2+}$. This ion is *mononuclear* just as the parent aqua ion. Other mononuclear products of hydrolysis, such as $[Co(NH_3)_4(OH)_2]^+$ are formed by deprotonation of additional water ligands. The loss of two protons from a single water ligand produces an Oxo-ligand such as in $[VO(NCS)_4(H_2O)]^{2-}$.

2. Polynuclear products of hydrolysis are formed by elimination of water molecules from the primary, mononuclear hydroxoaqua ions (olation). An example of olation is the formation of the binuclear di-μ-hydroxo complex ion $[(H_3N)_4Co(OH)_2Co(NH_3)_4]^{4+}$ by the solid state reaction

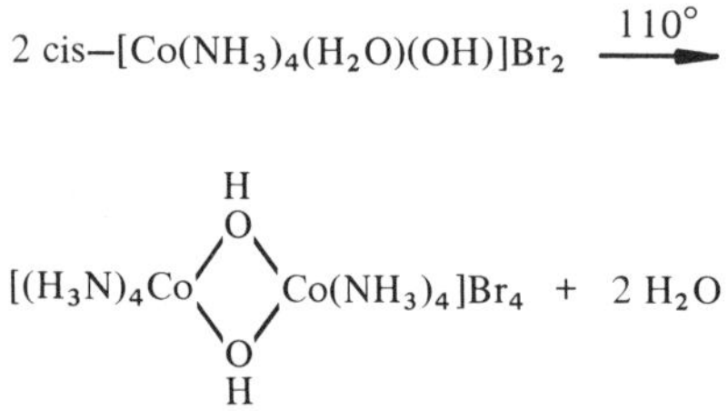

(2)

Similar condensation reactions take place in aqueous solution. One, two or three μ-(OH) bridges may be formed between two metal atoms. A μ-hydroxo ligand may lose its proton and form a μ-oxo bridged ion such as $[(H_3N)_5CrOCr(NH_3)_5]^{4+}$. Polynuclear species of high molecular weight may be formed by successive elimination of water molecules from hydroxoaqua ions. Most aged metal hydroxides, such as $Fe_2O_3 \cdot xH_2O$, belong to this group.

Two essential conclusions of this scheme of hydrolysis, which remained unchallenged since the beginning of the century are:

(a) The primary products of hydrolysis are *mononuclear* species in which a distinction can be made between an aqua-ligand and a hydroxo-ligand.

(b) In polynuclear products of hydrolysis, the bridging oxygen atom is bonded *directly* to, at least, two metal atoms.

These conclusions will be examined critically, in view of our recent findings concerning the nature and occurrence of the hydrogen-oxide bridging ligand.

The existence of the $H_3O_2^-$ ligand was demonstrated by single crystal x-ray studies of cluster ions of W(IV) and Mo(IV)[1–3] and of classical hydroxoaqua ions of Cr(III) and Co(III)[4, 6]. It was shown to exist also in a Ru(III) hydroxoaqua complex whose structure had been determined earlier[6, 12]. It is now proposed that this ligand exists in most hydroxoaqua metal ions. These ions, in the crystalline state, are not mononuclear, but binuclear or polynuclear. Furthermore, a distinction between a hydroxo-ligand and an aqua-ligand is impossible, if the proton forming the H-bond of $H_3O_2^-$ is equidistant from both oxygen atoms. An H_3O_2 bridge was also shown to exist in compounds which were previously mistaken for double-salts such as trans-$[Co(en)_2N_3(H_2O)] \cdot [Co(en)_2N_3(OH)](ClO_4)_3$. These salts are, in reality, not made up of aqua-ions and hydroxo-ions at a 1:1 ratio but of symmetrical binuclear ions having an H_3O_2 ligand between the metal atoms[7].

The existence of hydrogen-oxide bridging in hydroxoaqua ions in the crystalline state, raised the possibility that these hydrogen-oxide bridged species persist in *aqueous solution*. Primary results[5], obtained by Three Phase Vapor Tensiometry (TPVT), confirm the existence of binuclear hydroxoaqua ions in concentrated solutions. The existence of these ions throws new-light on the mechanism of some substitution and redox reactions of metal ions.

II. Very Strong O–H–O Bonds

The hydrogen bond of the $H_3O_2^-$ bridging ligand belongs to the class of "very strong" H-bonds. The subject of strong hydrogen bonds has been extensively reviewed in recent years[13–15]. The strength of a hydrogen bond A–H–B is most conveniently defined by the distance between the atoms A and B. If this distance is shorter than the sum of their van der Waals radii, by more than 0.5 Å, the bond is "very strong". A difference between 0.3 Å and 0.5 Å defines a "strong" H-bond, while <0.3 Å characterizes a "weak" H-bond[14]. For the homonuclear O–H–O bond, the corresponding separations are: smaller than 2.5 Å for a "very strong" bond, 2.5 to 2.7 Å for a "strong" bond and over 2.7 Å for a "weak" bond.

A decrease in the separation between the oxygen atoms is accompanied by a lowering of the *potential well* in which the hydrogen atom sits. The corresponding H-bond energies are: >100 KJ/mole for a "very strong" bond, >50 KJ/mole for a "strong" bond and <50 KJ/mole (often <30 KJ/mole) for a "weak" bond.

The O---O distance also affects the *shape* of the potential energy function of the proton. It is generally agreed[13–15] that the shortest and strongest H-bonds have a single-minimum halfway, or approximately halfway, between the oxygen atoms. This minimum becomes shallower as the separation increases and eventually splits into two minima separated by an energy barrier that becomes higher as the separation increases.

Single crystal-neutron diffraction is the best method for determining that location of the proton in the H-bond. X-ray diffraction is less precise by an order of magnitude, due to the small x-ray scattering power of the hydrogen atom.

In many strong H-bonds, the center of the bond coincides with a point of crystal symmetry. This is consistent with a proton being at the center of the bond, but it may also be the result of *two* H-atom sites on either side of the bond center (double-minima potential) which are related statistically or dynamically. A distinction between these two alternatives may sometimes be made by ir or nmr spectroscopy, but the results are often inconclusive[15]. If the center of the bond does not coincide with a point of symmetry, lattice forces may displace the hydrogen from the central position; it may be nearer to one oxygen atom and it may not be colinear with the oxygens, i.e., form an O–H–O angle of less than 180°. Some compounds containing very strong O–H–O bonds are presented in Table 1. Compounds **1** to **6** belong to the group of acid salts $MH(OA)_2$ in which the two negative OA^- ions are bonded by a very strong H-bond $[AO{-}H{-}OA]^-$. Compounds **7** and **8** contain the diaqua-hydrogen ion $[H_5O_2]^+$.

The diaqua-hydrogen ion $[(H_2O){-}H{-}(OH_2)]^+$ exists in acidic aqueous solutions; its very strong H-bond was characterized mainly by ir spectroscopy[16]. The ion is bonded to other solvent molecules by *weak* H-bonds, forming species such as $H_9O_4^+$. Discrete $H_5O_2^+$ ions exist in many crystals such as **8** and **9** in Table 1. The O---O separation is usually between 2.4 and 2.5 Å, but in one case a bond length of only 2.336 Å was reported[17].

In basic solution the hydrogen oxide anion $[HO{-}H{-}OH]^-$ was investigated by ir spectroscopy[16]. This ion too is extensively bonded to solvent molecules by weak H-bonds, forming species such as $H_7O_4^-$.

A discrete $H_3O_2^-$ anion with an unusually short O---O separation of 2.29(2) Å was reported in the mixed salt $Na_2[Et_3MeN]\{Cr[PhC(S){=}N(O)]_3\} \cdot \frac{1}{2} Na(H_3O_2) \cdot 18\ H_2O$[18].

Table 1. Some compounds with very strong H-bonds[a]

No.	Compound	Diffraction method	O—O Å	Character of H-bond
1	$NaH(MeCO_2)_2$	X-ray	2.444	symmetric
2	$KH(MeCO_2C_6H_4CO_2)_2$	neutron	2.448	H-centered
3	$KH(MeCH{=}CHCO_2)_2$	neutron	2.488	$O \overset{1.141\,Å}{—} H \overset{1.348\,Å}{—} O$
4	$N_2H_5^+HC_2O_4^-$	X-ray	2.450	symmetric
	$N_2H_5^+HC_2O_4^-$	neutron	2.448	H-centered
5	$RbH[O(CH_2CO_2)_2]$	neutron	2.449	$O \overset{1.226\,Å}{—} H \overset{1.226\,Å}{—} O$
6	$CsH(NO_3)_2$	neutron	2.468	$O \overset{1.236\,Å}{—} H \overset{1.236\,Å}{—} O$ 172.6°
7	$HBr \cdot 2\ H_2O$	neutron	2.40	$O \overset{1.17\,Å}{—} H \overset{1.22\,Å}{—} O$ 174.70°
8	trans-$[Co(en)_2Cl_2]Cl \cdot 2\ H_2O \cdot HCl$	neutron	2.431	centered

[a] Data taken from Ref. 13

The $H_3O_2^-$ unit is symmetric with a crystallographic inversion center between the oxygen atoms. Weak hydrogen bonds exist between the $H_3O_2^-$ ion and water molecules in the crystal. It is noteworthy that x-ray studies of hydrates of strong hydroxides such as $Cs(OH) \cdot H_2O$[19] and $(Me_4N)OH \cdot 5\,H_2O$[20] did *not* reveal discrete $H_3O_2^-$ ions but structures with infinite frameworks. In $Cs(OH) \cdot H_2O$ the high temperature structure is characterized by layers of $(H_3O_2^-)_n$ polyanions with an O---O distance of 2.64 Å. In $(Me_4N)(OH) \cdot 5\,H_2O$, there is a hydrogen bonded polyhedral framework of OH^- ions and H_2O molecules which are statistically disordered over the framework sites. The O---O distances in this structure are within the range of 2.735–2.794 Å.

III. $H_3O_2^-$ Bridging in Trinuclear Clusters

Preparation[21] and structural characterization[22] of quadruply bonded binuclear molybdenum(II) compounds with bridging carboxylate ligands was reported more than 20 years ago. These yellow compounds were obtained by reacting $Mo(CO)_6$ with the appropriate carboxylic acid. The reaction with acetic acid produces $Mo_2(OAc)_4$ in low yields (20–25%) and the fate of the rest of the molybdenum products remained unknown for quite some time. In 1976, it was reported that the dark brown solution remaining after the separation of the $Mo_2(OAc)_4$ crystals contained trinuclear, triangular clusters of molybdenum[23]. Subsequent work in several laboratories revealed a mixed collection of triangular cluster compounds having the general formula $[M_3X_2(O_2CR)_6L_3]^{n\pm}$ (X = O, CCH_3, CCH_2CH_3; R = CH_3, CH_2CH_3, $C(CH_3)_3$; L = O_2CCH_3, H_2O) all of which are obtained by reacting $M(CO)_6$ (M = Mo, W) with carboxylic acids[24].

In all these clusters, each metal atom has a coordination number of 9, counting the two neighboring metal atoms and the coordinated atoms of the various ligands. A general representation of the M_3X_{17} core of the trinuclear clusters is depicted in Fig. 2. One particular group of clusters having the formula $[M_3O_2(O_2CR)_6(H_2O)_3]^{2+}$ contains molybdenum or tungsten atoms with a formal oxidation state of +4 and two triply bridging capping oxygen atoms. Each pair of metal atoms is bridged by two carboxylate ligands

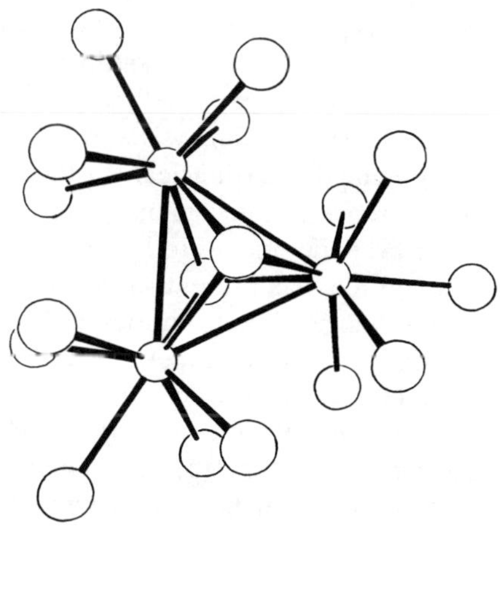

Fig. 2. A general representation of the M_3X_{17} unit

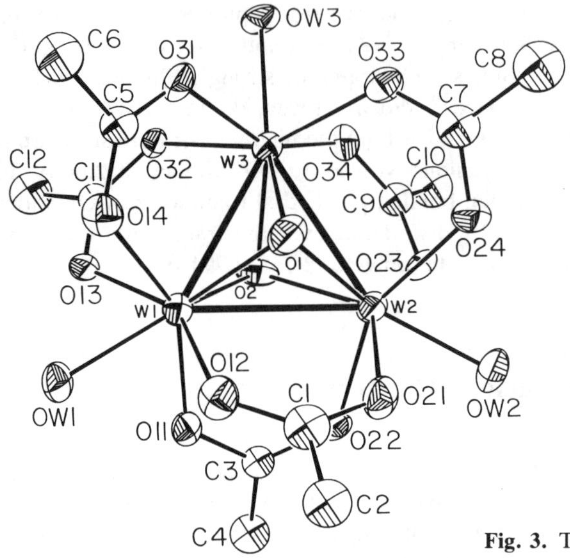

Fig. 3. The structure of $[W_3O_2(OAc)_6(H_2O)_3]^{2+}$

and each metal atom is coordinated to a water ligand (Fig. 3). The most prominent feature of these trinuclear species is their unusual kinetic stability in aqueous solution. All attempts to substitute ligands in the first coordination sphere of the metal cluster have failed. This stability is attributed to the high coordination number of the metal atoms in the cluster. These cluster compounds were usually prepared by elution of the 2+ cation from a cation exchange column with acids such as HBF_4, CF_3SO_3H and $HClO_4$, followed by slow evaporation of the eluate. Substances such as $[W_3O_2(OAc)_6(H_2O)_3](CF_3SO_3)_2$[24b] and $[W_3O_2(pr)_6(H_2O)_3](BF_4)_2$[24a] were obtained by this procedure and their structures were determined by x-ray structural analysis.

In 1980 an attempt was made in our laboratory to prepare the molybdenum analogue of the tritungsten hexapropionato cluster, i.e. $[Mo_3O_2(pr)_6(H_2O)_3]^{2+}$ in order to compare their structural features. The red ion was obtained almost quantitatively by reacting $Mo(CO)_6$ with propionic acid followed by addition of H_2O to the reaction mixture and adsorption on a cation exchange column. Due to a delay in delivery of HBF_4 to our laboratory, a solution of 0.5 M KBr was used for elution of the ion from the column. The red eluate, was placed in an open beaker for evaporation and yielded red crystals after several days. We expected these crystals to be $[Mo_3O_2(pr)_6(H_2O)_3]Br_2$ but results of the single crystal x-ray analysis were puzzling and unexpected[1]. The unit cell, of space group $P\bar{1}$ contained two trinuclear clusters. The existence of two 2+ cations in the cell required the presence of four anions, in order to balance the charge. To our great surprise only three bromide ions were located and there was no indication that any other negatively charged ion was present in the cell. The two trimers appeared in pairs in this crystal, rather than as discrete entities, as in related trinuclear clusters that had been investigated before[24]. The two trimers related to each other through a crystallographic center of symmetry which was located midway between two equatorial oxygen atoms, each coordinated to a metal atom of a different cluster. The short O---O distance of 2.52 Å, led us to the conclusions that: (a) a short *hydrogen bond* existed between these two atoms; and (b) that this hydrogen bond was centred, since the two M–O distances were equal (symmetry

Table 2. Structural data for $H_3O_2^-$ bridging ligands in trinuclear metal cluster compounds

No.	Compound	O–O(H_3O_2), Å	M–O(H_3O_2), Å	M···M, Å	M–O–O(H_3O_2), deg.	M–O···O–M torsional angle, deg.	Reference
9	$\{[Mo_3O_2(pr)_6(H_2O)_2]_2(H_3O_2)\}Br_3 \cdot 6\,H_2O$, dicluster	2.52(1)	2.009(7)	5.63	117.1	180	2
10	$\{[W_3O_2(pr)_6(H_2O)_2]_2(H_3O_2)\}Br_3 \cdot 6\,H_2O$, dicluster	2.50(1)	1.99(1)	5.64	119.4	180	2
11	$\{[W_3O_2(pr)_6(H_2O)_2]_2(H_3O_2)\}(NCS)_3 \cdot H_2O$, dicluster	2.46(1)	2.04(1)	5.73	120.3	180	3
12	$\{[Mo_3O_2(pr)_6(H_2O)_2]_2(H_3O_2)\}(NCS)_3 \cdot H_2O$, dicluster	2.52(1)	2.01(1)	5.71	119.9	180	3
13	$[W_3O_2(pr)_6(H_2O)(H_3O_2)]NCS$, polycluster	2.44(1)	2.02(1) 2.07(1)	5.95	132.9 130.0	156.1	3
14	$[W_3O_2(OAc)_6(H_2O)(H_3O_2)NCS$, polycluster	2.44(2)	2.06(1) 2.03(1)	5.81	126.5 126.2	149.6	3

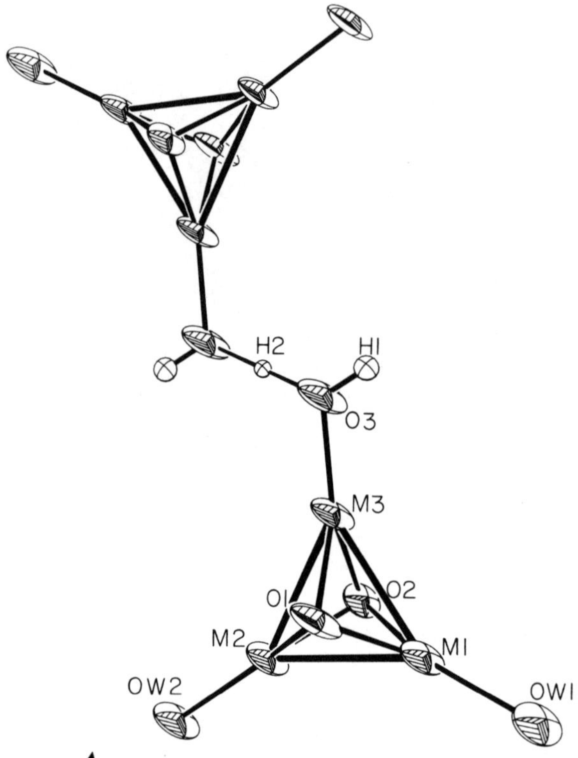

Fig. 4. Skeletal structure of $\{[M_3O_2(pr)_6(H_2O)_2]_2(H_3O_2)\}^{3+}$, dicluster. The propionato groups are omitted for the sake of clarity. Ref. 1

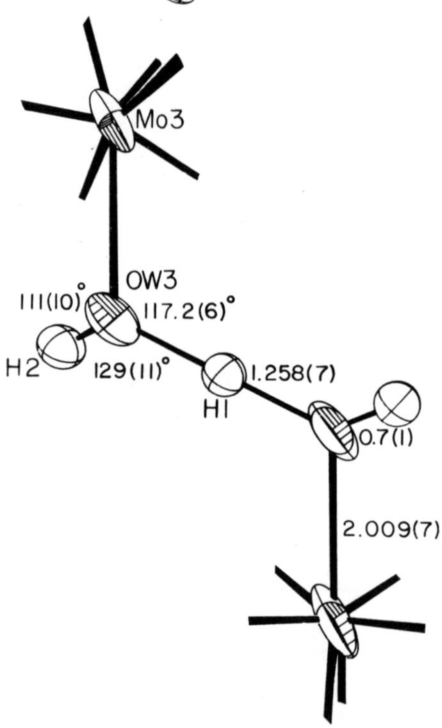

Fig. 5. The structure and dimensions of the Mo–O(H)HO(H)–Mo portion of the dicluster (the $H_3O_2^-$ bridging ligand). Ref. 2

related). The only acceptable explanation for this phenomenon was, that one of the coordinated water ligands lost a proton and that a new kind of *bridging ligand* was formed, the $H_3O_2^-$ bridging ligand. This $H_3O_2^-$ ligand is formed by a strong and symmetric hydrogen bond between an OH^- ligand of one cluster and an H_2O ligand of another. The presence of this uninegatively charged bridging entity explained the "missing" charge in the crystal. The compound was therefore formulated as $[Mo_3O_2(pr)_6(H_2O)_2$–H_3O_2–$Mo_3O_2(pr)_6(H_2O)_2]Br_3 \cdot 6\,H_2O$, **9**. Table 2 presents structural information of 9 and its tritungsten analogue **10** which was prepared by a similar procedure[2]. A closer examination of the M–M distances within the trimer showed that the M_3 triangles were not equilateral but isosceles. This distortion was a result of a displacement of the bridged metal atom towards the negatively charged $H_3O_2^-$ unit. In a more detailed x-ray structural analysis of **9**[2] the positions of the hydrogen atoms in the molecule, including those in the $H_3O_2^-$ bridge, were determined more accurately. The structure is shown in Fig. 4. Figure 5 shows the $H_3O_2^-$ unit.

The formation of a hydroxo ligand in the cluster (Eq. 3) is obviously the result of hydrolysis of the triaqua ion (the "monocluster")

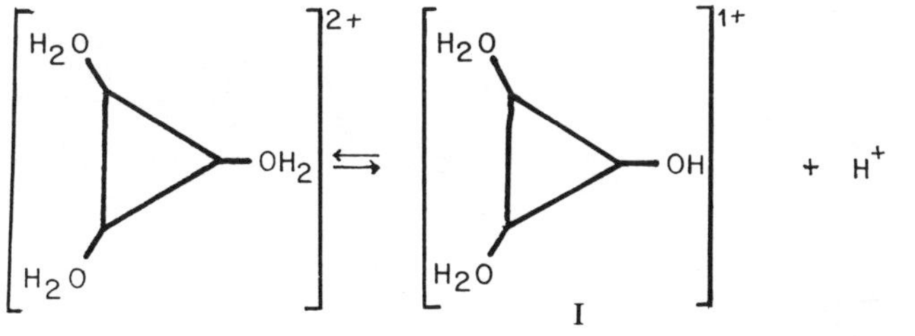

(3)

The hydroxo species I reacts with a monocluster to form a dicluster by means of the H_3O_2 bridging ligand (Eq. 4)

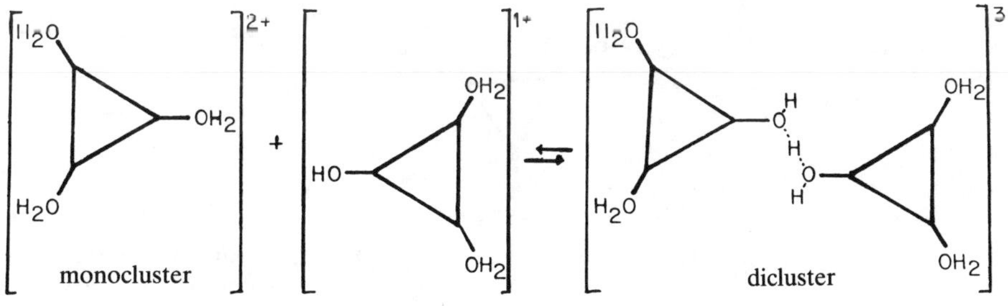

(4)

Similarly, the reaction of several I with each other may form a polymeric chain of clusters ("polycluster") as in reaction 5

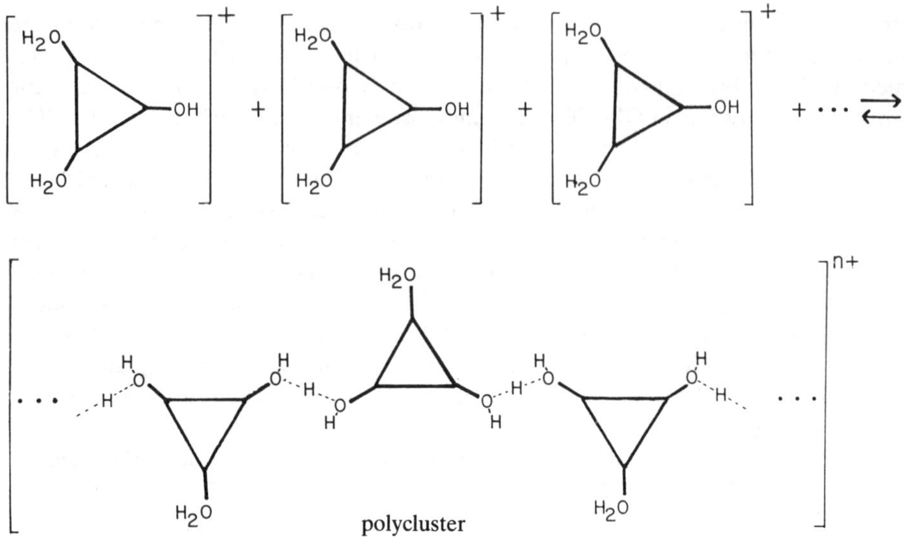

(5)

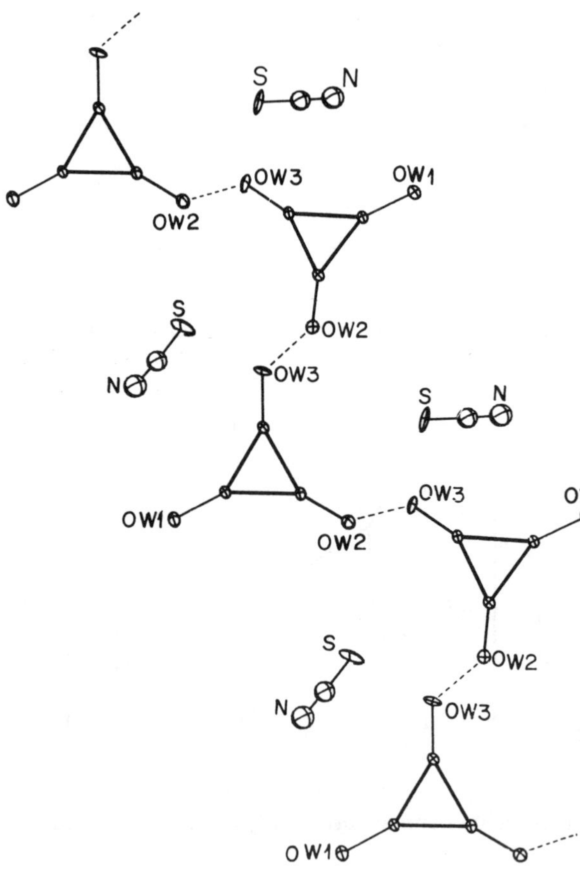

Fig. 6. Infinite chain of the polycluster in 13. The μ_3-oxygen atoms and propionate ligands were omitted for the sake of clarity. The O---H---O bonds in the $H_3O_2^-$ ligands are represented by *dotted lines*. Ref. 3

For a given total concentration of clusters, the relative concentration of these species depends on the H^+ concentration. At low pH, the concentration of the dicluster increases and reaches its maximum concentration at pH = pKa. At still higher pH the equilibrium in solution shifts towards the polycluster. The validity of this scheme was demonstrated[3] by varying the pH of solutions containing equimolar concentrations of $[W_3O_2(pr)_6(H_2O)_3]^{2+}$ in 0.5 M KNCS. At an H^+ concentration of 1 M crystals of the triaqua monocluster ion $[W_3O_2(pr)_6(H_2O)_3](NCS)_2$ **11** are obtained. At $[H^+] = 0.05$ M crystals of the dicluster $\{[W_3O_2(pr)_6(H_2O)_2]_2(H_3O_2)\}(NCS)_3 \cdot H_2O$ **12**, with bridging H_3O_2, are deposited. When no acid is added (pH ~ 4–5) crystals of the polycluster $[W_3O_2(pr)_2(H_2O)(H_3O_2)]_n(NCS)_n$ **13** are obtained. The structural features of the dicluster ion in **12** are similar to those of the dicluster ion in **10**. A crystallographic inversion center is located between the two oxygens of the $H_3O_2^-$ unit with an O---O (H_3O_2) separation of 2.46(1) Å[3]. The trinuclear clusters of the polycluster **13** form an infinite chain in the crystal, linked by $H_3O_2^-$ bridges as in Fig. 6. A similar polycluster of the hexaacetato complex $[W_3O_2(OAc)_6(H_2O)(H_3O_2)]_n(NCS)_n$, **14** was prepared in an analogous way[25] and its structure is shown in Fig. 7. The O---O (H_3O_2) separation in both polyclusters is 2.44(1) Å.

A further increase in the pH causes deprotonation of a second water ligand of the monocluster:

$$[Mo_3O_2(O_2CR)_6(H_2O)_2(OH)]^+ \rightleftharpoons \underset{\text{II}}{[Mo_3O_2(O_2CR)_6(H_2O)(OH)_2]} + H^+ \quad (6)$$

Crystals of II were obtained by different preparative routes[26].

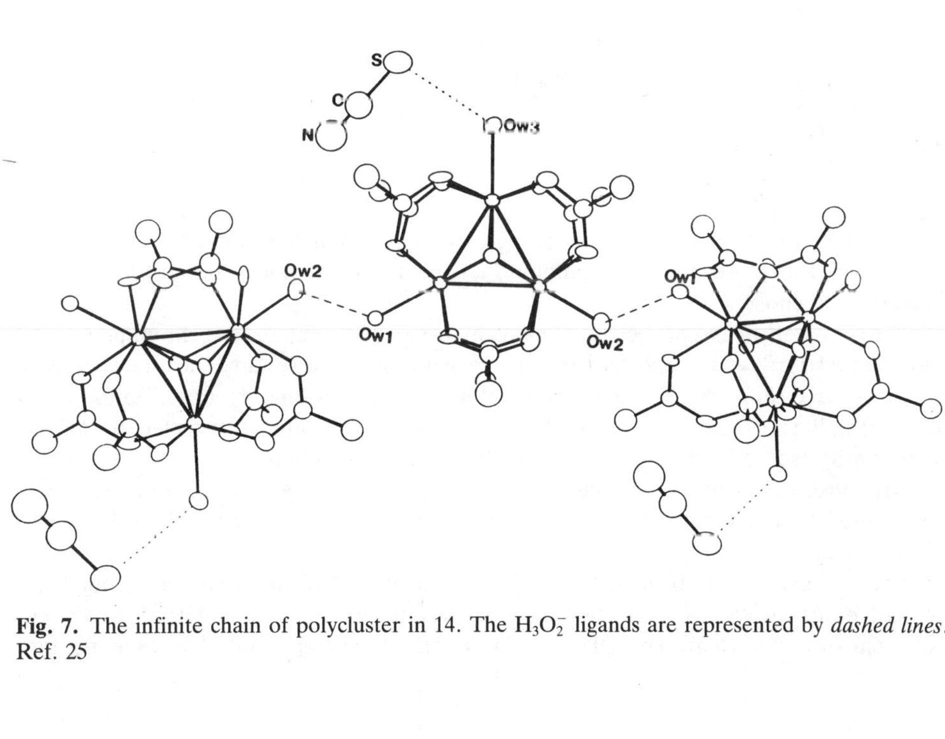

Fig. 7. The infinite chain of polycluster in 14. The $H_3O_2^-$ ligands are represented by *dashed lines*. Ref. 25

IV. Hydroxoaqua Metal Ions

The structural investigation of classical hydroxoaqua metal ions was triggered by the realization that metal atoms of trinuclear clusters do not have any unique property that would make them more susceptible to hydrogen-oxide bridging than ordinary, mononuclear metal ions. This conclusion led to a working hypothesis that H_3O_2 bridging may be a fundamental, hitherto unrecognized, property of many metal ions having H_2O and OH^- ligands.

The strategy employed in the search of evidence for this hypothesis was to grow single crystals of hydroxoaqua complexes of chromium(III) and cobalt(III) and to determine their structure by x-ray crystallography. The choice of these metals was not accidental. They were chosen because of their inertness to substitution reactions. Hydrogen oxide bridged species such as cis-$[L_4M(H_3O_2)_2ML_4]^{(2n-2)+}$ could be obtained by a very rapid reaction of H-bond formation

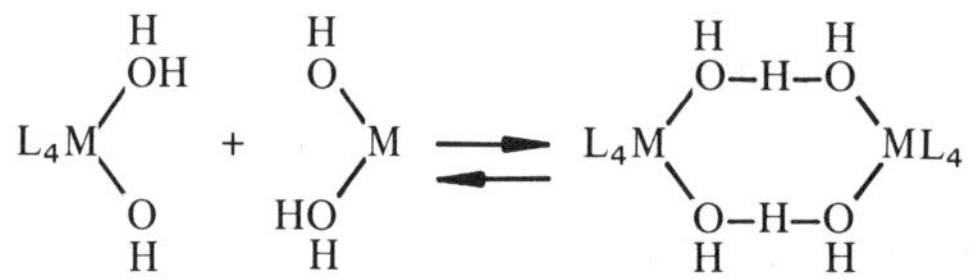
(7)

However the product of reaction (7) is expected to be a short-lived intermediate, the end product being the di-μ(OH) ion which is thermodynamically more stable:

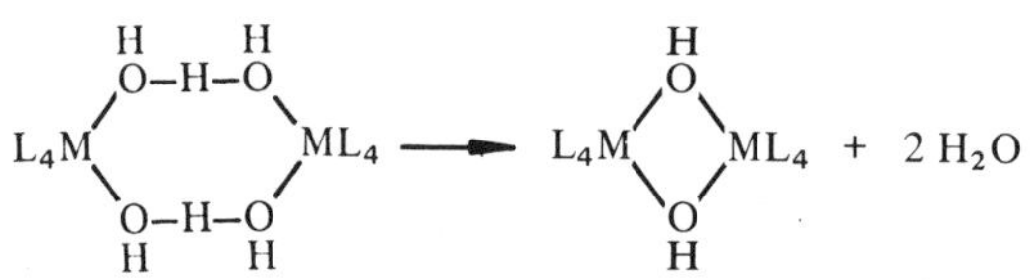
(8)

Reaction (8) which involves the breaking of metal-oxygen bonds is slow only in inert complexes. Therefore the most common inert complexes, namely those of Cr(III) and Co(III) were chosen.

The preparation and properties of many hydroxoaqua complexes of chromium(III)[27a)] and cobalt(III)[27b)] were reported since the beginning of the century, but some of them cannot be used for growing single crystals. Rapid crystallization or recrystallization yields microcrystalline precipitates while slow crystallization is often obstructed by olation (reaction 8), isomerization and other slow substitution reactions.

Structures of seven compounds of hydroxoaqua complexes were determined so far. Some important structural features of these compounds are presented in Table 3 and Figs. 8 to 12.

These structures confirm our assumption that many hydroxoaqua metal complexes are not mononuclear as their classical formulation implies (compounds **15** to **19**) but binuclear or polynuclear. The OH^- and H_2O ligands in compounds **15** to **20** merge into

Table 3. Structural data for $H_3O_2^-$ bridging ligands in octahedral metal complexes

No.	Compound	O–O(H_3O_2), Å	M–O(H_3O_2), Å	M···M, Å	M–O–O(H_3O_2), deg.	M–O···O–M torsional angle, deg.	Reference
15	cis-$[Cr(bpy)_2(H_3O_2)]_2I_4 \cdot 2\,H_2O$	2.446(5)	1.925(3) 1.928(3)	5.03	127.1(2) 126.2(2)	64.9	6
16	$[Co(tren)(H_3O_2)]_2(NO_3)_4 \cdot 2\,H_2O$	2.450(4)	1.916(2) 1.924(2)	4.82	118.7(1) 125.9(1)	64.5	28
17	cis-$[Cr(bpy)_2(H_3O_2)]_2(NO_3)_4$	2.442(4)	1.934(3) 1.913(4)	4.96	128.7(2) 125.6(2)	54.0	28
18	trans-$[Co(en)_2(H_3O_2)](ClO_4)_2$	2.441(2)	1.916(1)	5.72	130.4(1)	180	6
19	trans-$[Ru(bpy)_2(H_3O_2)](ClO_4)_2$	2.538(6)	2.007(3)	5.79	127.3(1)	137.7	12
20	cis-$\{[Ir(en)_2]_2(OH)(H_3O_2)\}$ $(S_2O_6)_{3/2} \cdot ClO_4 \cdot 2.75\,H_2O$	2.429(9)	2.070(7) 2.074(6)	3.80	111.0(3) 107.4(3)	7.4	29
21	cis-$[Cr(C_6H_8N_2)(H_2O)(OH)]S_2O_6$[a]	2.586(6)	1.998(5)(H_2O) 1.926(4)(OH)	6.18	141.2 144.3	166.1	30
22	trans-$\{[Co(en)_2(NO_2)]_2(H_3O_2)\}$ $(ClO_4)_3 \cdot 2\,H_2O$	2.412(9)	1.906(6)	5.67	129.8(3)	180	7
23	trans-$\{[Co(en)_2(NCS)]_2(H_3O_2)\}$ $(CF_3SO_3)_3 \cdot H_2O$	2.415(6)	1.911(5) 1.916(5)	5.77	135.9(3) 132.3(3)	174.3	7
24	trans-$\{[Cr(NH_3)_3(OH)]_2(H_3O_2)\} \cdot Br_3 \cdot 2\,H_2O$	2.450(5)	1.983(3) 1.975(4)	5.64	123.5(1) 125.4(1)	147.2	39

[a] This is a genuine hydroxo-aqua complex without $H_3O_2^-$ ligands (see text)

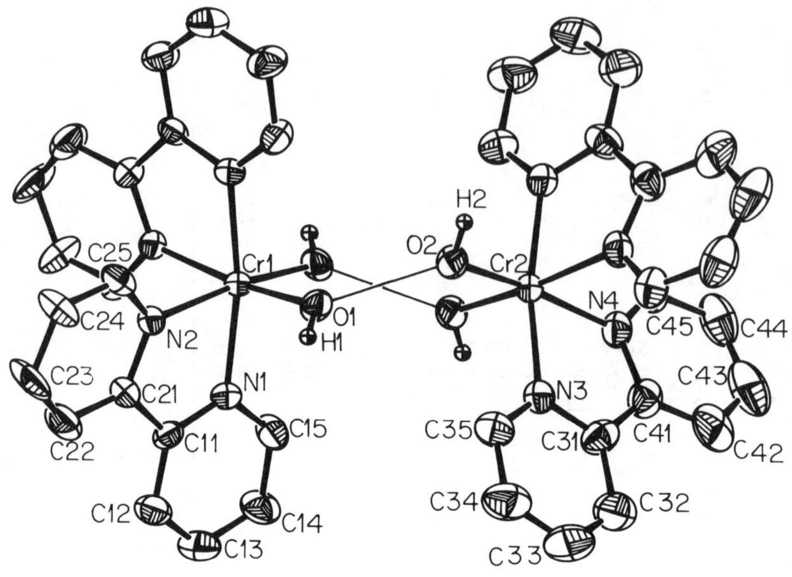

Fig. 8. Structure of $[(bpy)_2Cr(H_3O_2)_2Cr(bpy)_2]^{4+}$. Ref. 6

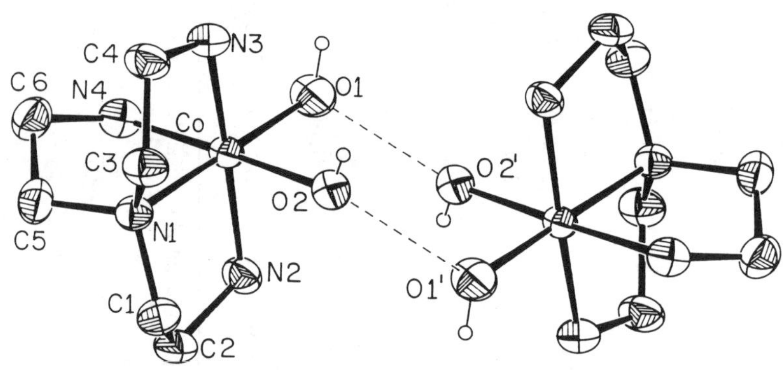

Fig. 9. The structure of $[(tren)Co(H_3O_2)_2Co(tren)]^{4+}$. Ref. 6

an $H_3O_2^-$ bridging ligand by means of a very short hydrogen bond, 2.42 Å to 2.45 Å (except in the ruthenium compound **19** in which it is 2.538 Å).

Of the seven compounds only one (compound **21**) contains distinguishable hydroxo and aqua ligands that do not merge into H_3O_2 bridges. Therefore only this compound is a genuine hydroxoaqua complex. In all other compounds, no meaningful distinction between these ligands is possible because the hydrogen bond that joins them, in the H_3O_2 ligand, is centered or approximately centered.

The centered position of the proton in the H_3O_2 bridges of compounds **15** to **20** is confirmed by the length of the two metal-oxygen bonds: A bond M---O(OH) between a metal atom and a hydroxo-ligand is considerably shorter than the M---O(H_2O) bond between the same metal and an aqua-ligand. In chromium(III), for example,

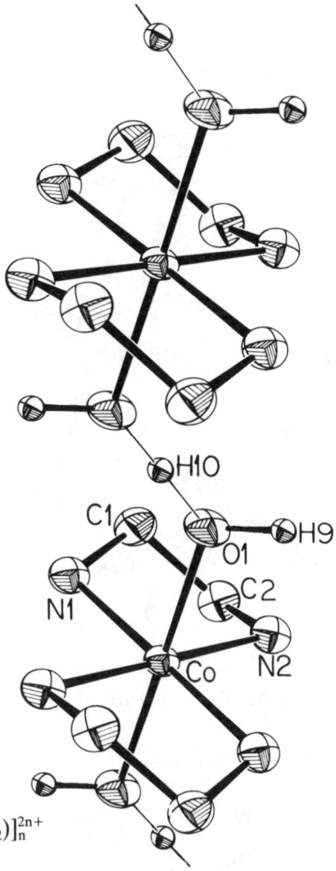

Fig. 10. Section of the infinite chain of trans-$[(Co(en)_2)(H_3O_2)]_n^{2n+}$

Cr---O(OH) in most compounds is in the range of 1.90–1.91 Å[31] and Cr---$O(H_2O)$ in the range of 1.98–2.01 Å[32]. This appreciable difference makes possible a distinction between genuine hydroxo and aqua ligands. In compound **21** one Cr---O separation is 1.998(5) Å while the other is 1.926(1) Å. The first is clearly a Cr–OH_2 bond and the second is a Cr–OH bond. It is remarkable that the moderately strong H-bond that joins the oxygen ligands of neighboring chromium atoms in compound **21** [2.586(6) Å] has only a very slight effect on the two Cr–O bond lengths. This H bond is definitely *not* centered, the hydrogen being covalently bonded to the aqua-oxygen and hydrogen-bonded to the hydroxo-oxygen. In contrast to this exceptional situation in compound **21**, one finds in all other compounds (**15** to **24**) two metal oxygen bonds that are of equal or nearly equal length. Therefore the very strong hydrogen bond between the oxygen atoms in these compounds must be centered or approximately centered and no meaningful distinction can be made between H_2O and OH^- ligands. These compounds are neither aqua nor hydroxo-complexes. Their M---$O(H_3O_2)$ separation is shorter than a genuine M---$O(H_2O)$ bond but longer than a M---O(OH) bond.

Compound **15** to **17** have binuclear cations in which the two metal atoms are bridged by *two* H_3O_2 ligands as in Fig. 8 and Fig. 9. Such double bridging is possible only if the two oxygen ligands are in *cis* positions. In compound **21**, which is also a cis isomer, the formation of this double bridge is blocked by the dithionate counterion (Fig. 12). The

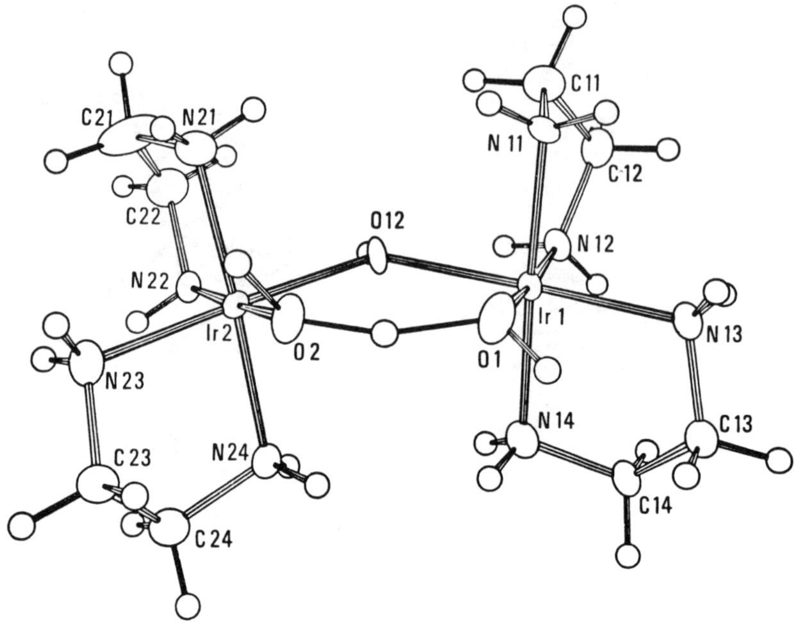

Fig. 11. The structure of cis-$\{[Ir(en)_2]_2(OH)(H_3O_2)\}^{4+}$. Ref. 29

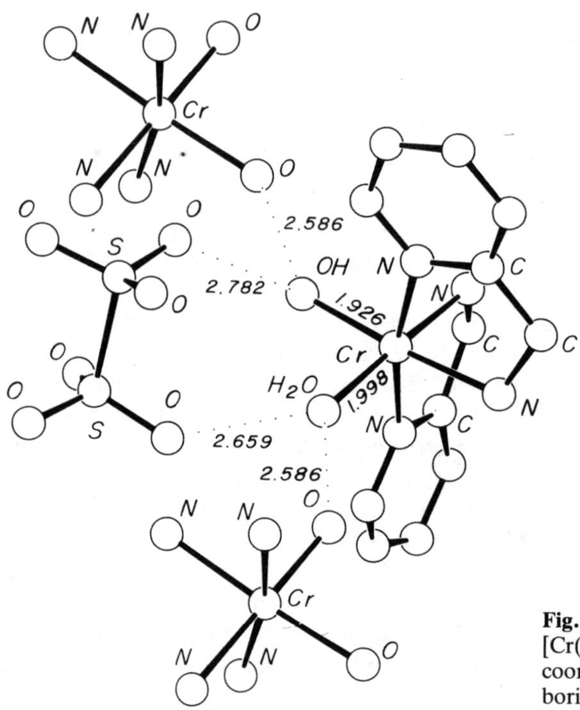

Fig. 12. The structure of cis-$[Cr(C_6H_8N_2)(H_2O)(OH)]S_2O_6$. Only first coordination spheres are drawn for neighboring complexes

trans complexes **18** and **19** have polynuclear chains of metal atoms which are linked by single H_3O_2 bridges as in Fig. 10. Inspection of Fig. 8 and Fig. 9 reveals that there are two forms of double bridging by H_3O_2. The two O–O axes may be parallel as in **16** and **17** or not, as in **15**. Another way to describe this difference is by characterizing the six membered ring of these compounds (not counting the central H-atom)

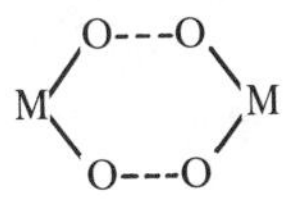

This ring has a "chair" configuration in **16** and **17** and a "skew" configuration in **15**. These configurations are probably determined by packing forces and not by the nature of the metal atom or its ligands. This point is illustrated by compounds **15** and **17** that have the same $[Cr(bpy)_2(H_3O_2)]_2^{4+}$ cation in a skew and chair configuration respectively.

The spatial relation of the two M–O bonds of the H_3O_2 bridge may be characterized as "gauche" (Fig. 13a) in compounds **15** to **17** with a torsional angle of approximately 60° and as "anti" in compound **18** with a torsional angle of 180° (Fig. 13b). In compound 20 which has one μ-(OH) and one μ-(H_3O_2) bridge between the two iridium atoms, the torsional angle is only 7.4°, which is almost that of a cis configuration (0°). These different configurations may be rationalized by the following considerations. Repulsion between the positively charged metal atoms tends to increase the separation between them. In a singly bridged structure, this separation may attain a maximum, at 180°, as in compound **18**[1]. The double bridge geometry of compounds **15** to **17** does not permit an "anti" structure and imposes a "gauche" arrangement instead. The configuration of compound **20** is governed by the short μ-(OH) bridge which draws the metal atoms much closer (3.80 Å) than the longer H_3O_2 bridge. Therefore a cis configuration of the latter is favored in this compound. Furthermore, in order to accommodate this H_3O_2 bridge the IrOO angles (107.4° and 111.0°) have to be smaller than in the other compounds (127.1° and 126.2° in **15** and 130.4° in **18**).

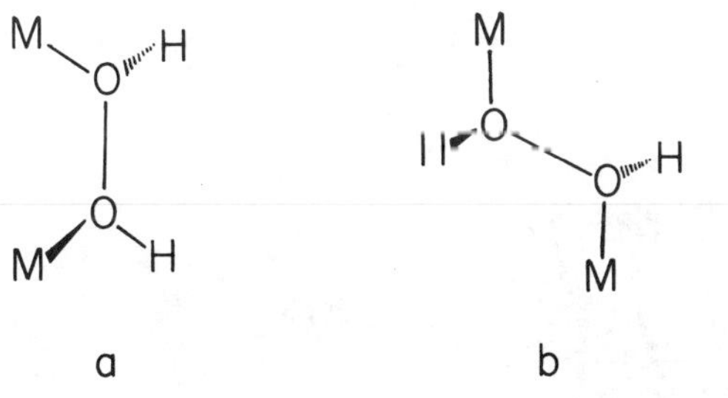

Fig. 13. **(a)** gauche configuration. **(b)** anti configuration

1 Packing forces and other factors may modify this angle as in compound **19** where it is only 138°

V. Aqua-Hydroxo Double Salts

Neutralization of a *monoaqua* ion such as $[ML_5(H_2O)]^{n+}$ by one equivalent of OH^- produces the classical, mononuclear hydroxo ion $[ML_5(OH)]^{(n-1)+}$. If less than one equivalent of OH^- is added, a mixture of monoaqua and monohydroxo ions is obtained. At pH = pK the ratio of these two species is of course 1:1. When such a 1:1 solution is concentrated, either the salt of the aqua ion or that of the hydroxo ion will crystallize, depending on their relative solubilities. However, in some cases neither the aqua salt nor the hydroxo salt but an "aqua hydroxo double salt" such as $[Co(en)_2(OH_2)N_3]$-$[Co(en)_2(OH)N_3](ClO_4)_3$ will precipitate[33]. Results summarized in Sects. III and IV raised the possibility that these so-called double salts may in reality consist of symmetric binuclear ions bridged by a single H_3O_2 ligand. The structures of two such "double salts" were determined[7]: trans-$\{[Co(en)_2(NO_2)]_2(H_3O_2)\}(ClO_4)_3 \cdot 2\,H_2O$, **22** (Fig. 14), and trans-$\{[Co(en)_2(NCS)]_2(H_3O_2)\}(CF_3SO_3)_3 \cdot H_2O$, **23** (Fig. 15).

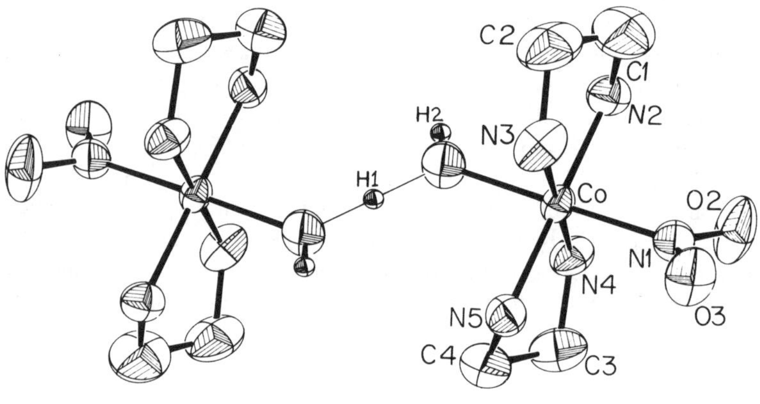

Fig. 14. The structure of trans-$\{[Co(en)_2(NO_2)]_2(H_3O_2)\}^{3+}$

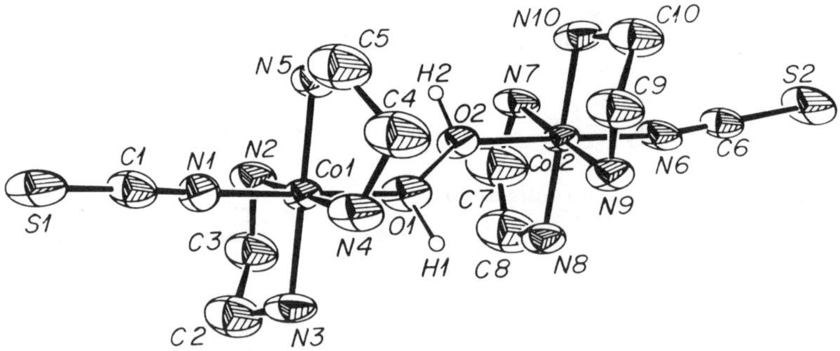

Fig. 15. The structure of trans-$\{[Co(en)_2(NCS)]_2(H_3O_2)\}^{3+}$

A double salt formulation for compound **22**, $[Co(en)_2(NO_2)(OH_2)][Co(en)_2(NO_2)(OH)](ClO_4)_3 \cdot 2\,H_2O$ implies the existence of *two* distinct cationic species in the crystal, an aqua ion and a hydroxo ion, having two different Co–O distances. Structures **22** and **23** are not of this kind. They are definitely not double salts, but salts of a single unique, binuclear cation $[(en)_2XCo(H_3O_2)CoX(en)_2]^{3+}$ ($X = NO_2$, NCS). The two cobalt atoms in this cation have the same coordination sphere. They have neither the features of a hydroxo ion nor those of an aqua ion. The Co–O(H_3O_2) distance, is intermediate between those found in hydroxo and aqua ions. In **22** the two M–O(H_3O_2) distances of 1.906(6) Å are symmetry related whereas in **23** they are identical within the experimental error – 1.911(5) Å and 1.916(5) Å (Table 1). These distances are shorter than the average distance reported for the Co–O(H_2O) bond (1.94 Å)[34] but longer than the Co–O(OH) distance (1.89 Å)[35]. Due to the limitation of the x-ray technique, the exact position of the central hydrogen atom in the H_3O_2 bridge cannot be located. However, the fact that the two M–O(H_3O_2) distances in both structures are identical supports the conclusion that the H-bond of the H_3O_2 ligand is centered.

The O---O distances in the H_3O_2 bridges of **22** and **23**, 2.412(9) Å and 2.415(6) Å respectively, are significantly shorter than those found in other compounds with H_3O_2 bridges (2.43–2.54 Å). The decrease in the O---O separation is probably the result of an inductive effect of the electron withdrawing NO_2^- and NCS^- ligands in the trans position. From Table 3 it can be seen that the torsional angle of the M–O---O–M system in **22** and **23** is 180° and 174.3(4)° respectively. This arrangement yields the longest M–M distance and the smallest interactions between the two positively charged metal centers, as in compound **19**.

The binuclear H_3O_2 bridged structure found in these compounds may be expected in other, so-called, double salts because the strong H-bond of the H_3O_2 bridge (≈ 100 KJ) makes this structure more stable than a hypothetical structure with distinct mononuclear hydroxo and aqua ions. However, before the general role of H_3O_2 bridging in coordination chemistry was outlined[6], a double salt formulation such as $[Co(en)_2(NO_2)OH] \cdot [Co(en)_2(NO_2)H_2O](ClO_4)_3$ seemed to be more justified than a symmetric formulation such as $[Co(en)_2(NO_2)(OH)]_2(ClO_4)_2 \cdot HClO_4$ which would imply the existence of an acid salt of a hydroxo-complex. The authors were therefore pleasantly surprised to find a paper by Alfred Werner (1907), which proposed an essentially correct formulation for this class of compounds, that was later mistaken for double salts[36]. This paper, bearing the title "Über anomale anorganische Oxonium Salze, eine neue Klasse basischer Salze" (on anomalous oxonium salts, a new class of basic salts) reports the composition of basic salts of trans-$[Co(NH_3)_4(NO_2)H_2O]^{2+}$. The structure proposed by Werner for the chloride salt is reproduced in Fig. 16. Werner's idea that a hydrogen atom may be coordinated with two (or more) oxygen atoms was proposed in an even earlier paper (1902) in which he first used the idea of a symmetric hydrogen bond (without using this term explicitly)[37]. It is noteworthy that Werner applied the idea of hydrogen bonding to structural problems as early as 1902, long before it was first put forward by Latimer and Rodebush as an explanation of the physical properties of water and related liquids[38].

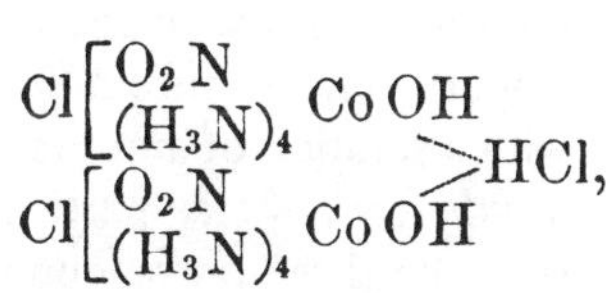

Fig. 16. Werner's formulation of trans-$\{[Co(NH_3)_4(NO_2)]_2(H_3O_2)\}Cl_3$[37a]

A recently reported structure of $[(H_2O)(NH_3)_3Cr(OH)_2Cr(NH_3)_3(OH)]Br_3 \cdot 2\,H_2O$ (compound **24**) was described as being built of an infinite, hydrogen-bonded, chain of alternating trans-diaqua and trans-dihydroxo binuclear ions[39]. This characterization, which fits a double salt, is not supported by the crystallographic data (Table 3). The H-bond in this chain is very strong (2.45 Å) and the two Cr–O bonds on each side of it are of nearly equal length, as in other H_3O_2 bridged species. Therefore compound **24** should be formulated as trans-$[Cr(NH_3)_3(OH)]_2(H_3O_2)Br_3 \cdot 2\,H_2O$ and it consists of chains of *chemically* (though not crystallographically) identical units bridged by H_3O_2 ligands ... $(H_3O_2)[(NH_3)_3Cr(OH)_2Cr(NH_3)_3](H_3O_2)[(NH_3)_3Cr(OH)_2Cr(NH_3)_3]$ ...

VI. Hydroxoaqua Metal Ions in Solution

The existence of H_3O_2 bridges between metal atoms in the crystalline state raised the possibility that these bridges persist in aqueous solution of hydroxoaqua metal ions. This possibility seemed reasonable, since breaking these bridges requires approximately 100 KJ/mol for each bridge, a factor which may not be cancelled by the entropy increase accompanying this process (especially in high concentrations). If equilibria such as

$$2[(H_2O)_5Fe(OH)]^{2+} \rightleftharpoons [(H_2O)_4Fe(H_3O_2)_2Fe(H_2O)_4]^{4+} \tag{9}$$

exist in solution, they may affect, not only hydrolytic processes **per-se** but also thermodynamic and mechanistic features of other reactions of metal ions in solution.

The dimerization of cis-$[Cr(bpy)_2(H_2O)(OH)]^{2+}$ in solution was studied by titration of the corresponding diaqua ion cis-$[Cr(bpy)_2(H_2O)_2]^{3+}$ with OH^-. This titration converts the diaqua ion, first into the hydroxoaqua ion and then into the dihydroxo ion cis-$[Cr(bpy)_2(OH)_2]^+$. If dimerization of the hydroxoaqua ion occurs, these reactions should be formulated as follows

$$[Cr(bpy)_2(H_2O)_2]^{3+} + OH^- \rightleftharpoons \tfrac{1}{2}[(bpy)_2Cr(H_3O_2)_2Cr(bpy)_2]^{4+} + H_2O \tag{10}$$

$$\tfrac{1}{2}[(bpy)_2Cr(H_3O_2)_2Cr(bpy)_2]^{4+} + OH^- \rightleftharpoons [Cr(bpy)_2(OH)_2]^+ + H_2O \tag{11}$$

Evidence for dimerization of the hydroxoaqua ion is based on the *decrease* of the number of chromium particles in reaction 10 and its increase in reaction 11. The number, ν, of discrete chromium particles per diaqua ion should decrease from $\nu = 1$ to $\nu = \frac{1}{2}$ as the diaqua ion is titrated with 1 mol of OH^- and then rise again to $\nu = 1$ as a second mole of OH^- is added. ν may be determined by measuring a colligative property of the solution. A most suitable method for ionic solutes is Three-Phase Vapor Tensiometry, TPVT[40,41]. The three-phase solvent system consists of a saturated solution of an electrolyte in water, in equilibrium with the crystalline phase of that electrolyte and with water vapor. An isobaric temperature difference $(\Delta T)_p$ is established when the pure solvent is equilibrated with a solution of a foreign solute in the same solvent, at constant pressure. The *apparent number,* ν_m of free particles per formula of solute depends on the molality of the solute (m), the three-phase ebulioscopic constant K_e, and $(\Delta T)_p$

$$\nu_m = (\Delta T)_p/K_e m$$

The apparent particle number ν_m depends linearly on m.

$$\nu_m = \nu + Nm$$

ν, the true number of free particles per formula of solute, is determined by linear extrapolation to m = 0. The validity of this relationship was confirmed experimentally in over 40 solute-solvent systems[40)], using solutes with known ν. The precision of ν determinations is ± 0.05.

The *solute* used was $[Cr(bpy)_2(H_2O)_2](NO_3)_3 \cdot \frac{1}{2} H_2O$[5)]. A saturated solution of barium nitrate was used as solvent. A weighed sample of $[Cr(bpy)_2(H_2O)_2](NO_3)_3 \cdot \frac{1}{2} H_2O$ was dissolved in the saturated barium nitrate solution and neutralized *stepwise* by the addition of weighed portions of $Ba(OH)_2 \cdot 8\, H_2O$. ν_m was determined after the addition of each portion of barium hydroxide. Since both the nitrate counterion of the chromic species and the barium ion added in the form of barium hydroxide were common to solute and solvent, they did not affect ν_m, which was exclusively dependent on the chromic species. If dimerization of $[Cr(dpy)_2H_2O(OH)]^{2+}$ occurred, the initial ν_m of the diaqua salt solution should be *reduced* by the addition of barium hydroxide and reach its minimum value after 1 equiv of barium hydroxide was added. The results, presented in Fig. 6 confirm this expectation. Addition of more than 1 equiv of OH^- converts the dinuclear hydroxoaqua ion to the mononuclear dihydroxo ion and raises ν_m as expected. However, dimerization is not complete at a molality of 0.24 m (formula weights of the

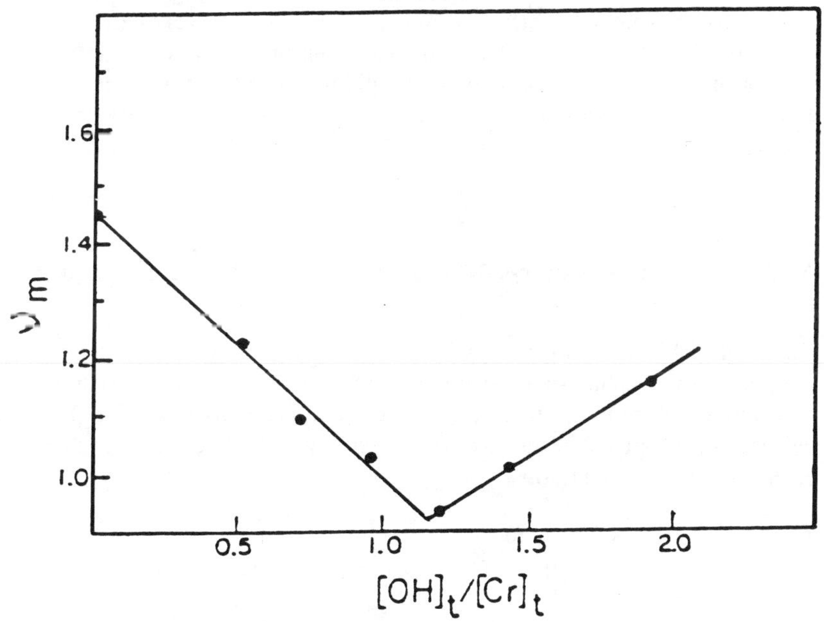

Fig. 17. TPVT titration of a solution of $[Cr(bpy)_2(H_2O)_2](NO_3)_3 \cdot \frac{1}{2} H_2O$ in a saturated solution of barium nitrate at 25 °C. Molality of total chromium species in mol Cr per 1 kg water = 0.24 m. Ref. 5

diaqua salt), since ν_m is not reduced to 50% of its initial value, but only to $\approx 64\%$ (from 1.45 to 0.92). This indicates that an equilibrium is established in solution between the mononuclear and the dinuclear hydroxoaqua ions

$$2\,[Cr(bpy)_2H_2O(OH)]^{2+} \rightleftharpoons [(bpy)_2Cr(H_3O_2)_2Cr(bpy)_2]^{4+} \quad (12)$$

TPVT titrations of cis-$[Co(en)_2(H_2O)_2]^{3+}$ and $[Co(tren)(H_2O)_2]^{3+}$ gave similar results[42]. The equilibrium constant of (12) was approximately 1 in these three ions, so that in dilute solutions ($< 10^{-2}$ M) the tetrapositive, binuclear ions will be almost completely dissociated into dipositive hydroxoaqua ions. The last conclusion was confirmed[43] by Feltham-Onsager plots[44] of the equivalent conductance (at 0 °C) versus the square-root of the concentration of four salts: $[Co(NH_3)_4(H_2O)(OH)](NO_3)_2$; $Cr(NH_3)_4(H_2O)(OH)](NO_3)_2$; $[Co(NH_3)_4(H_2O)(OH)]Br_2$[45] and $[Cr(H_2O)_6](NO_3)_3$ + NaOH[43]. The conductivity data were for solutions of concentrations between 10^{-2} and 10^{-3} M. At these concentrations the Feltham-Onsager plots are linear[44]. All four plots had the typical slope of 1 : 2 electrolytes which was less than half of the slope of 1 : 4 electrolytes.

A TPVT study was also conducted on the "double salt" 22[42]. The results indicated that at a concentration of ~ 0.1 M the ion $[Co(en)_2NO_2]_2(H_3O_2)^{3+}$ was completely dissociated into an aqua and a hydroxo ion. The higher stability of a doubly bridged dimer compared to a singly bridged binuclear ion is the result of higher bond energy of the former (two strong H-bonds compared to one) and the chelate effect.

The formation of a hydrogen oxide bridging ligand in the hydrolysis of the mononuclear iron(III) complex, $[Fe(EHGS)(H_2O)]$ in solution, was recently supported by Mossbauer spectroscopy[46]. At low and high pH, the Mossbauer spectra are characteristic of mononuclear species (the aqua and the hydroxo complexes respectively) with well defined paramagnetic hyperfine structure. At pH $\sim$ pKa, very narrow absorption lines of a fast relaxing iron site are observed, indicating the formation of a dimer. The appearance of the fast-relaxing component at a pH in which the concentrations of the hydroxo and aqua complexes are equal suggest that the dimers are formed by means of $H_3O_2^-$ bridging.

VII. Mechanism of Reactions Involving $H_3O_2^-$ Bridging

One of the objects of this review is to initiate new studies concerning the role of $H_3O_2^-$ bridging in the mechanism of reactions of metal ions. Since no such studies have yet been reported, this section will be confined to discussion and reinterpretation of some reaction mechanisms proposed in the past, in view of the recently established, existence of $H_3O_2^-$ bridging between metal ions.

VII. 1. Mechanism of Olation Reactions

Reactions of hydroxoaqua complexes such as

$$2\,[Cr(en)_2(H_2O)(OH)]X_2 \rightarrow [(en)_2Cr(OH)_2Cr(en)_2]X_4 + 2\,H_2O \quad (13)$$

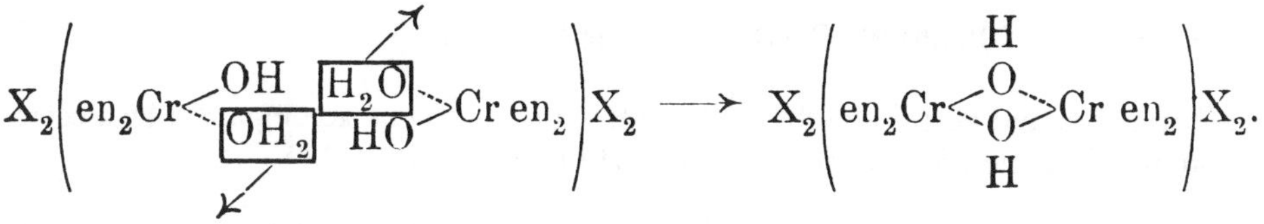

Fig. 18. Pfeiffer's scheme of the olation mechanism[47)]

were investigated since the beginning of this century. Pfeiffer noted that only *cis* isomers react at 100 °C while the trans isomers are stable at the same temperature. He proposed a scheme for this reaction which is reproduced from his original paper[47)] in Fig. 18. Pfeiffer comments that the two molecules are pushed towards each other in such a way that a hydroxo ligand of one molecule substitutes a water ligand of the other molecule and splits it off the complex. This scheme requires a rearrangement of the hydroxoaqua ions in pairs, in which the oxygen ligands of one chromium are facing those of the other chromium atom. The three structures of cis complexes **15**, **16** and **17** (Figs. 8 and 9, Table 3) reveal that such a rearrangement is unnecessary since it already exists in the *stable* configuration of the reactant: the complex is binuclear and the oxygen ligands of the two metal atoms face each other and are held in place by the strong H-bonds of the H_3O_2 bridges.

Olation reactions also take place in aqueous solution. A well known example is the olation of the Cr(III) aqua ion

$$2\,[Cr(H_2O)_5OH]^{2+} \rightleftharpoons \underset{\text{III}}{[(H_2O)_4Cr(OH)_2Cr(H_2O)_4]^{4+}} + 2\,H_2O \tag{14}$$

which was shown to produce the di-μ-hydroxo ion III[48)]. The olation reaction is completely reversed in acid solution:

$$[L_4M(OH)_2ML_4]^{4+} + 2\,H^+ + 2\,H_2O \rightarrow 2\,[ML_4(H_2O)_2]^{3+} \tag{15}$$

The existence of an equilibrium between monomeric hydroxoaqua ions and dimeric H_3O_2 bridged ions in solution:

$$2\,[ML_4(H_2O)(OH)]^{(n-1)+} \rightleftharpoons [ML_4(H_3O_2)]_2^{2(n-1)+} \tag{16}$$

offers an explanation to the puzzling kinetics of the olation reaction

$$2\,\text{cis-}[Cr(C_2O_4)_2(H_2O)(OH)]^{2-} \rightarrow [(C_2O_4)_2Cr(OH)_2Cr(C_2O_4)_2]^{4-} + 2\,H_2O \tag{17}$$

This is a first order reaction in concentrated solutions but changes to second order in dilute solutions[49)]. If this reaction proceeds via the dimer $[(C_2O_4)_2Cr(H_3O_2)_2Cr(C_2O_4)_2]^{4-}$ this change of order follows directly from the effect of concentration on the equilibrium of reaction 16.

It was shown that reaction 15 proceeds by two consecutive steps[50, 51)].

$$H_2O + [L_4M(OH)_2ML_4]^{4+} \rightleftharpoons \underset{\text{IV}}{[(H_2O)L_4M(OH)ML_4(OH)]^{4+}} \tag{18}$$

$$H_2O + 2\,H^+ + [(H_2O)L_4M(OH)ML_4(OH)]^{4+} \rightleftharpoons 2\,[ML_4(H_2O)_2]^{3+} \tag{19}$$

The formation and destruction of the two OH bridges may be accompanied by a corresponding destruction and formation of two H_3O_2 bridges

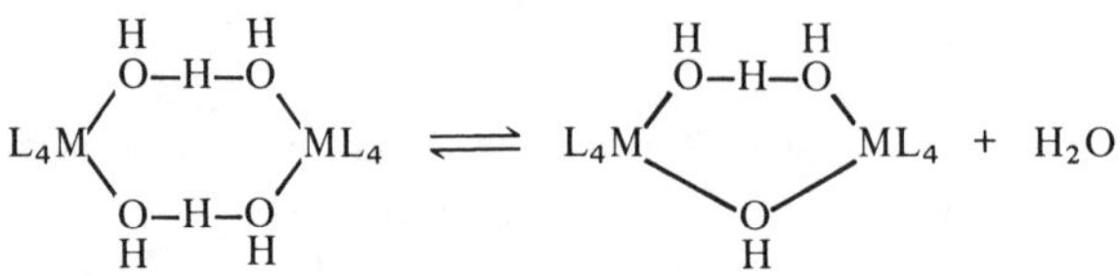

V IV

(20)

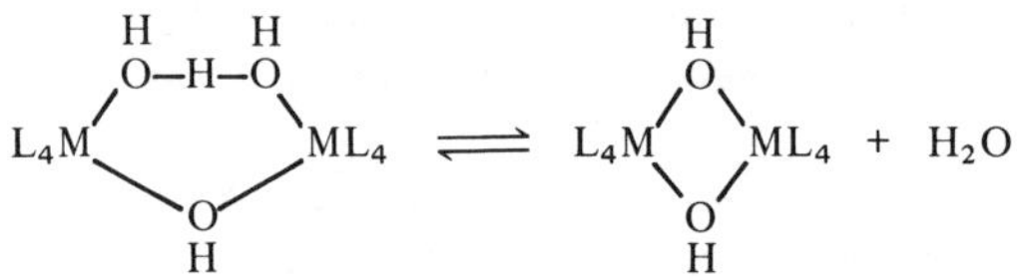

(21)

Springborg and coworkers[52)] investigated the thermodynamics and kinetics of the equilibration reaction 18 and isolated salts of the stable intermediate $[(H_2O)L_4M(OH)ML_4(OH)]^{4+}$. They found indirect evidence for the existence of a hydrogen bond between the terminal aqua and hydroxo ligand of IV and have recently proved its existence in a crystal of compound **20**[29)]. This compound (Fig. 11, Table 3) does, in fact, have an intramolecular $H_3O_2^-$ bridge with a symmetric H-bond and a characteristic O---O distance of 2.429 Å. As noted in Sect. IV of this review, considerable strain is imposed on this ligand by the shorter μ-OH bridge. As a result of this strain, the H_3O_2 configuration is nearly cis, instead of gauche and the IrOO angles are smaller than usual. The fact that reaction 21 is always faster than 20[52)] may be rationalized by the lower stability of the strained H_3O_2 bridge of IV compared to that of a regular H_3O_2 bridge of V.

Solid state olation rejections, such as 13, may proceed by two steps similar to those taking place in solution (Eqs. 18 and 19). One may account for the reactivity of the cis hydroxoaqua complexes and the inertness of the trans isomers by comparing their structures. In order to form one OH bridge in the cis compound **15** (Fig. 8), one H_3O_2 bridge has to be broken. But formation of a similar OH bridge in the trans compound 18 (Fig. 10) requires the breaking of an *additional* H_3O_2 bridge of the infinite chain, in order to let the two metal atoms decrease their distance from ~5.7 Å to ~3.8 Å. Therefore the activation energy for olation of a trans-isomer is expected to be much higher and would prevent the reaction under comparable conditions.

The rapid formation of $H_3O_2^-$ bridges between hydroxo and aqua ligands may play a role in the complex and slow process of *aging* of metal hydroxides. This process, which is accompanied by a decrease of solubility in acids, involves olation reactions which lead to the formation of polynuclear frameworks of metal atoms bridged by hydroxo and oxo bridges. The first step of precipitation of metal aqua ions by OH^- may be the formation of a loosely bound structure in which the metal atoms are bridged by $H_3O_2^-$ ligands. These bridges collapse during the aging process and are replaced by shorter OH^- bridges. This mechanism is illustrated by the case of Chromium(III) hydroxide. The freshly precipitated $Cr(OH)_3 \cdot 3\,H_2O$ does not contain OH bridges, because it dissolves in acid without producing any polynuclear ions. If the precipitate is allowed to age, OH bridges are formed, and persist after dissolution in acid. The freshly precipitated hydroxide is microcrystalline and is converted by aging to an amorphous substance[8, 53]. This aging process is too quick to allow growing of single crystals of $Cr(OH)_3 \cdot 3\ H_2O$ for x-ray analysis. The x-ray powder patterns of this substance suggest an inverse *bayerite* type structure in which the chromium atoms occupy vacant octahedral sites. The isolated CrO_6 octahedra are arranged in the hexagonal crystal in layers within the ab planes and held together by strong hydrogen bonds. Since all Cr–O bonds are crystallographically identical and no distinction can be made between Cr–OH and Cr–OH_2 it is reasonable to assume that this compound contains chromium(III) atoms, bridged by symmetrical $H_3O_2^-$ units. Therefore, the formula of the chromium(III) hydroxide should be $[Cr(H_3O_2)_3]$.

VII. 2. Redox Reactions

Over thirty years ago an H atoms transfer mechanism was proposed by Silverman and Dodson[54] for the Fe(III)/Fe(II) exchange reaction

$$[Fe^*(H_2O)_5(OH)]^{2+} + [Fe(H_2O)_6]^{2+} \rightleftharpoons [Fe^*(H_2O)_6]^{2+} + [Fe(H_2O)_5(OH)]^{2+} \tag{22}$$

Such a H atom transfer may be carried out by cleavage of an H_3O_2 ligand bridging the oxidizing and reducing ions:

$$[(H_2O)_5Fe^*(OH)]^{2+} + [(H_2O)Fe(H_2O)_5]^{2+} \rightleftharpoons [(H_2O)_5Fe^*\text{–}\overset{\text{H}}{\text{O}}\text{–H–}\overset{\text{H}}{\text{O}}\text{–}Fe(H_2O)_5]^{4+} \tag{23}$$

$$[(H_2O)_5Fe^*\text{–}\overset{\text{H}}{\text{O}}\text{–H–}\overset{\text{H}}{\text{O}}\text{–}Fe(H_2O)_5]^{4+} \rightleftharpoons [(H_2O)Fe^*(H_2O)_5]^{2+} + [(HO)Fe(H_2O)_5]^{2+} + \tag{24}$$

Even if the concentration of the monobridged "mixed valence" dimer is very low, it may furnish a convenient reaction path if the H-bond of the H_3O_2 ligand is centered. There is, as yet, no experimental evidence supporting such a hydrogen atom transfer mechanism and no mixed valence compounds were reported in which two atoms of a metal in different oxidation states are bridged by H_3O_2 ligands.

VIII. Conclusions and Perspectives

The existence and widespread occurrence of H_3O_2 bridging between metal ions in crystals has been firmly established. More accurate data on the location of the central H atom may be obtained by neutron diffraction. The occurrence of H_3O_2 bridging in solution has been demonstrated by one experimental method (TPVT). Further support for these findings may be obtained by diffraction studies of solutions of hydroxoaqua complexes. Such studies may confirm the existence of H_3O_2 bridged dimers with a characteristic distance of approximately 5Å between the metal atoms. Elucidation of the role of H_3O_2 bridging in the mechanism of reactions may contribute to the understanding of some substitution and redox reactions, including enzymatic reactions in which an H atom transfer via H_3O_2 bridging may take place.

Acknowledgment. We are grateful to the Fund of Basic Research administered by the Israel Academy of Sciences and Humanities for financial support.

IX. Note Added in Proof

Since the manuscript of this review was submitted for publication in the Spring of 1985, further investigations of the structure and properties of complexes containing the H_3O_2 bridging ligand were carried out. Some revealed new and unexpected properties while others confirmed assumptions put forward in preceding sections of this review.

It was established that in some doubly-bridged, dimeric, cishydroxoaqua complexes of chromium(III) there is a low temperature diamagnetic interaction between the two chromium(III) atoms[55]. It is remarkable that this effect exists despite the long Cr...Cr distance (~5 Å) and is almost as strong as that found in the corresponding diols, which have a much smaller separation (~3 Å)[31, 32].

The mechanism of olation in the solid state, proposed in Sect. VII, was confirmed: the fact that only cis-hydroxoaqua complexes olate was ascribed to their doubly-bridged dimeric structure. A direct conclusion from this hypothesis is that any hydroxoaqua complex that does *not* have such a dimeric structure, should *not* olate. This conclusion was confirmed in two cases: the only complex of this type which was known not to have a dinuclear structure (compound 21, Table 3, Ref. 29), does not olate even at 200 °C and is decomposed at higher temperatures[56a]. When the dythionate counter ion in this salt was replaced by iodide, a normal, double-bridged dimer was obtained[56b].

This salt looses the calculated weight of water when heated to 140 °C for one hour and the product was shown to be a μ-diol by a single crystal X-ray study[56b]. Another cis-hydroxoaqua complex, α-cis-[Cr(bispicen)(H_3O_2)]I_2 (bispicen = N,N′-bis(2-pyridil-methyl)-1,2-ethanediamine) was found not to have a dimeric structure but an infinite chain of Cr(III) atoms linked by single H_3O_2 ligands. This compound too does not olate[57], thus reconfirming the hypothesis on the relation between olation and the double-bridged dimeric structure.

New evidence for the persistence of H_3O_2 bridging in concentrated solution, corroborating our earlier evidence[5], was obtained by a Differential anomalous x-ray scattering study of solutions of a trinuclear tungsten cluster[58].

X. References

1. Bino, A., Gibson, D.: J. Am. Chem. Soc. *103,* 6741 (1981)
2. Bino, A., Gibson, D.: ibid. *104,* 4383 (1982)
3. Bino, A., Gibson, D.: Inorg. Chem. *23,* 109 (1984)
4. Ardon, M., Bino, A.: J. Am. Chem. Soc. *105,* 7748 (1983)
5. Ardon, M., Magyar, B.: ibid. *106,* 3359 (1984)
6. Ardon, M., Bino, A.: Inorg. Chem. *24,* 1343 (1985)
7. Ardon, M., Bino, A., Jackson, W. G.: Polyhedron, *6,* 181 (1987)

8. Meyenburg, U., Siroky, O., Schwarzenbach, G.: Helv. Chim. Acta *56,* 1099 (1973)
9. Rohlfing, C. L., Allen, L. C., Cook, M. C.: J. Chem. Phys. *78,* 2498 (1983)
10. Roos, B. O., Kraemer, W. P., Diercksen, G. H. F.: Theor. Chim. Acta *42,* 77 (1976)
11. Werner, A.: Chem. Ber. *40,* 272 (1907); *40,* 2103 (1907); *40,* 4434 (1907); *40,* 4834 (1907)
12. Durham, B., Wilson, S. R., Hodgson, D. J., Mayer, T. J.: J. Am. Chem. Soc. *102,* 600 (1980)
13. Emsley, J.: Chem. Soc. Revs. *9,* 91 (1980)
14. Emsley, J., Jones, D. J., Lucas, J.: Reviews in Inorganic Chemistry *3,* 105 (1981)
15. Hadzi, D.: J. Molecular Structure *100,* 393 (1983)
16. Zundel, G.: The Hydrogen Bond, Recent Developments in Theory and Experiment. Structure and Spectroscopy (Shuster, P., Zundel, G., Sandorfy, C., Eds.) Vol. II, Chapter 15, Amsterdam, North Holland Publishing Co. 1979
17. Bino, A., Cotton, F. A.: J. Am. Chem. Soc. *101,* 4150 (1979)
18. (a) Abu-Dari, K., Raymond, K. N., Freyberg, D. P.: ibid. *101,* 3688 (1979)
    (b) Abu-Dari, K., Freyberg, D. P., Raymond, K. N.: Inorg. Chem. *18,* 2427 (1979)
19. Jacobs, H., Harbrecht, B., Muller, P., Bronger, W.: Z. Anorg. allg. Chem. *491,* 154 (1982)
20. McMullan, R. K., Mak, T. C. W., Jeffrey, G. A.: J. Chem. Phys. *44,* 2338 (1966)
21. Stephenson, T. A., Bannister, E., Wilkinson, G.: J. Chem. Soc. 2538 (1964)
22. (a) Lawton, D., Mason, R.: J. Am. Chem. Soc. *87,* 921 (1965)
    (b) Cotton, F. A., Mester, Z. C., Webb, T. R.: Acta Crystallgr. *B30,* 2768 (1974)
23. Bino, A., Ardon, M., Maor, I., Kaftory, M., Dori, Z.: J. Am. Chem. Soc. *98,* 7093 (1976)
24. (a) Bino, A., Cotton, F. A., Dori, Z., Koch, S., Küppers, H., Millar, M., Sekutowski, J. C.: Inorg. Chem. *17,* 3245 (1978)
    (b) Bino, A., Hesse, K. F., Küppers, H.: Acta Crystallogr. *B36,* 723 (1980)
    (c) Bino, A., Cotton, F. A., Dori, Z.: J. Am. Chem. Soc. *103,* 243 (1981)
    (d) Bino, A., Cotton, F. A., Dori, Z., Kolthammer, B. W. S.: ibid. *103,* 5779 (1981)
    (e) Ardon, M., Bino, A., Cotton, F. A., Dori, Z., Kaftory, M., Kolthammer, B. W. S., Kapon, M., Reisner, G. M.: Inorg. Chem. *20,* 4083 (1981)
    (f) Bino, A., Gibson, D.: Inorg. Chim. Acta *65,* L37 (1982)
    (g) Bino, A., Cotton, F. A., Dori, Z., Falvello, L. R., Reisner, G. M.: Inorg. Chem. *21,* 3750 (1982)
    (h) Ardon, M., Bino, A., Cotton, F. A., Dori, Z., Kaftory, M., Reisner, G. M.: ibid. *21,* 1912 (1982)
25. Bino, A., Gibson, D.: Inorg. Chim. Acta *101,* 29 (1985)
26. (a) Birnbaum, A., Cotton, F. A., Dori, Z., Marler, D. O., Reisner, G. M., Schwortzer, W., Skaice, M.: Inorg. Chem. 2723, (1983)
    (b) Bino, A., Gibson, D.: Inorg. Chim. Acta *104,* 155 (1985)
27. (a) Gmelins Handbuch der Anorganischen Chemie, System 52 "Chrom" part C, 8th edition, Verlag Chemie GmbH, Weinheim 1965
    (b) ibid. System 58 "Cobalt" part B, 8th edition, Verlag Chemie GmbH, Berlin 1930
28. Ardon, M., Bino, A.: Unpublished results
29. Galsbøl, F., Larsen, S., Rasmussen, B., Springborg, J.: Inorg. Chem. *25,* 290 (1986)
30. Larsen, S., Nielsen, K. B., Trabjerg, I.: Acta Chem. Scand. *A37,* 833 (1983)
31. (a) Kaas, K.: Acta Crystallogr. *B35,* 1603 (1979)
    (b) Hodgson, D. J., Pedersen, E.: Inorg. Chem. *19,* 3116 (1980)
32. (a) Chesick, J. P., Doany, F.: Acta Crystallogr. *B37,* 1076 (1981)
    (b) Cline, S. J., Kallesøe, S., Pedersen, E., Hodgson, D. J.: Inorg. Chem. *18,* 796 (1979)
33. Jackson, W. G., Sargeson, A. M.: Inorg. Chem. *17,* 1348 (1978)
34. Kanazawa, Y., Matsumoto, T.: Acta Crystallogr. *B32,* 282 (1976)
35. Kucharski, E. S., Skelton, B. W., White, A. H.: Aust. J. Chem. *31,* 47 (1978)
36. Werner, A.: Ber. *40,* 4122 (1907)
37. Werner, A.: Liebig. Annal. *322,* 296 (1902)
38. Latimer, W. M., Rodebush, W. H.: J. Am. Chem. Soc. *42,* 1419 (1920)
39. Andersen, P., Nielsen, K. M., Petersen, A.: Acta Chem. Scand. *A38,* 593 (1984)
40. (a) Magyar, B.: Helv. Chim. Acta *48,* 1259 (1965)
    (b) Ibid. *51,* 194 (1968)
    (c) Structure Bond. *14,* 111 (1973)
41. Magyar, B., Schwarzenbach, G.: Acta Chem. Scand. *A32,* 943 (1978)
42. Ardon, M., Magyar, B.: Unpublished results

43. Ardon, M.: Unpublished results
44. Feltham, R. D.: Inorg. Chem. *3,* 1038 (1964)
45. (a) King, H. J. S.: J. Chem. Soc. 1275 (1932)
(b) 520 (1933)
46. Carrano, C. J., Spartalian, K.: Inorg. Chem. *23,* 1993 (1984)
47. Pfeiffer, P.: Z. anorg. Chem. *56,* 261 (1907)
48. (a) Ardon, M., Plane, R. A.: J. Am. Chem. Soc. *81,* 3197 (1959)
(b) Stunzi, H., Rotzinger, F. P., Marty, W.: Inorg. Chem. *23,* 2160 (1984)
49. Grant, D. M., Hamm, R. E.: J. Am. Chem. Soc. *78,* 3006 (1956)
50. Hoffmann, A. B., Taube, H.: Inorg. Chem. *7,* 903 (1968)
51. Thompson, M., Connick, R. E.: ibid. *20,* 2279 (1981)
52. (a) Springborg, J., Toftlund, H.: Acta Chem. Scand. *A30,* 171 (1976)
(b) Christensson, F., Springborg, J.: ibid. *36A,* 21 (1982)
(c) Hancock, M., Nielsen, B., Springborg, J.: ibid. *36A,* 313 (1982)
53. (a) Giovanoli, R., Stadelmann, W., Feitknecht, W.: Helv. Chim. Acta *56,* 839 (1973)
(b) Giovanoli, R., Stadelmann, W., Gamsjager, H.: Chimia *27,* 170 (1973)
54. Silverman, J., Dodson, R. W.: J. Phys. Chem. *56,* 866 (1952)
55. Ardon, M., Bino, A., Michelsen, K., Pedersen, E.: Submitted for publication
56. (a) Michelsen, K.: Acta Chem. Scand., *A26,* 1517–1526 (1972)
(b) Ardon, M., Bino, A., Michelsen, K.: J. Am. Chem. Soc., *108* (1987)
57. Ardon, M., Bino, A., Michelsen, K., Pedersen, E.: To be published
58. Lorentz, R. D., Bino, A., Penner-Hahn, J. E.: J. Am. Chem. Soc. *108,* 8116–8117 (1986)

Some Structural Problems Examined Using the Method of Moments

Jeremy K. Burdett

Department of Chemistry and James Franck Institute, The University of Chicago, Chicago, Illinois 60637, USA

The method of moments, a technique used by solid state physicists over the past fifteen years or so, is used to develop a model to understand the global structural preferences of molecules and solids as a function of electron count. The model, of direct use to chemists, has ties to graph theory and its conclusions may be used to view structures in a topological sense by considering the variation in stability of structural fragments with electron polulation. The results cross the traditional borders of chemical endeavor. It is shown how several existing ideas, associated with the domains of organic, inorganic, organometallic and solid-state chemistry, may be phrased in terms of the same fundamental concept using these ideas.

I. Introduction . . . 30

II. The Electronic Energy . . . 31
A. The Electronic Density of States . . . 31
B. Generation of Moments . . . 33
C. Energies as a Function of Electronic Configuration . . . 35

III. Some Structural Problems . . . 38
A. Graphite and Related Nets . . . 38
B. The Crystal Structures of Tl, Pb, Bi and Po . . . 40
C. The Geometry of Allene . . . 43
D. Linear and Bent AH_2 Molecules . . . 44
E. Cis and Trans Transition Metal Dioxo Compounds . . . 46
F. The Rutile and α-PbO_2 Structures . . . 47
G. Busy, Idle and Metriotic Orbitals . . . 48
H. Boron and Carbon Structures . . . 52
I. Jahn-Teller Distortions in Octahedral Complexes . . . 55
J. The Jahn-Teller Effect in Cyclobutadiene and Other Polyenes . . . 59
K. The Peierls Distortion . . . 60
L. The Coloring Problem . . . 62
M. Three Membered Rings in Molecules . . . 70
N. The Woodward-Hoffmann Rules . . . 71
O. Expansion of the Secular Determinant . . . 76

IV. Some More Difficult Structural Problems . . . 77
A. The Crystal Structures of the Transition Metals . . . 77
B. Bond Angle Changes . . . 81
C. The Coordination Number Problem . . . 86

V. Conclusion . . . 87

VI. References . . . 89

I. Introduction

Chemists interested in the structures and properties of molecules have long relied upon the ideas of molecular orbital theory to provide them with models with which to understand a diverse range of observations. Today we have available good computational methods which enable us to reproduce numerically, for small systems, equilibrium internuclear separations and geometry and many other observable properties of these molecules, such as heats of reaction and activation energies. We also have a collection of orbital ideas based on simple concepts such as symmetry, overlap and electronegativity which are of enormous use in providing a global picture of electronic structure[1]. The principle of orbital symmetry conservation, via the Woodward-Hoffmann rules, for example, has had a tremendous influence on the way chemists think about reaction pathways. A particularly effective device too has been the use of the fragment orbital method – the building up of the orbitals of a complex molecule from the frontier orbitals of suitably chosen fragments. In this article we describe a method which gives us another tool to further our progress in understanding the diverse collection of molecules and solids which nature has provided. We will only describe structural problems of direct interest to chemists here and will deliberately omit topics of interest to the physicist or surface scientist.

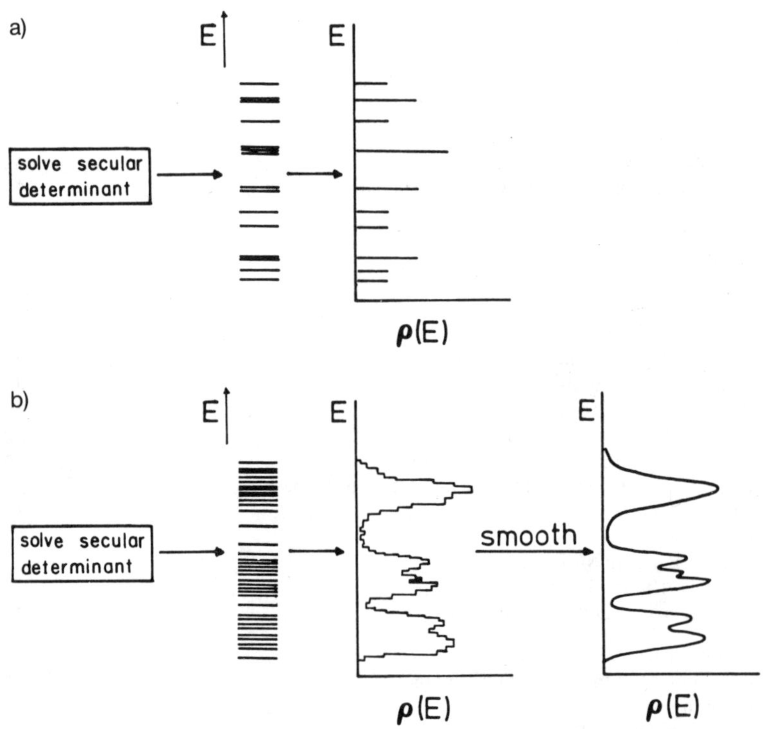

Fig. 1 a, b. Schematics showing the generation of the density of states for **a)** molecules and **b)** solids via the traditional route of determining the energy eigenvalues followed by construction of a density of states diagram

II. The Electronic Energy

A. *The Electronic Density of States*

Figure 1 shows schematically how the electronic density of states for a molecule or solid is conventionally obtained. Within the umbrella of Extended Hückel theory, which we use for its didactic simplicity, solution of the secular determinant

$$|\mathbf{H}_{ij} - \mathbf{S}_{ij}\mathbf{e}| = 0 \tag{1}$$

leads to a set of molecular orbital energy levels (eigenvalues) and orbital coefficients (eigenvectors), which may usefully be represented as a density of states diagram shown at the far right of Fig. 1a. The basis orbitals used in Eq. 1 are just the atomic orbitals of the molecule. If $\mathbf{S}_{ij} = \delta_{ij}$ then of course the original Hückel model results. The situation is a little more complex for solids since now there are an infinite collection of energy levels contained within the first Brillouin zone of the solid. In practice[2)] the secular determinant of Eq. 2 is

$$|\mathbf{H}_{ij}(k) - \mathbf{S}_{ij}(k)\mathbf{e}| = 0 \tag{2}$$

solved at a representative number of "k-points" within the zone. The basis set used in this calculation comprises the Bloch sums (Eq. 3).

$$\varphi_i(k) = \sum_l \exp(ik \cdot R_{li})\chi_i(r - R_{li}) \tag{3}$$

k is the wavevector and the $\{\chi_i\}$ the orbital contents of the unit cell. The sum is over all unit cells of the crystal, l, where orbital χ_i is located on atoms residing at R_{li} with respect to some arbitrary origin. The finite collection of levels obtained by multiple solution of Eq. 2 are collected into a density of states plot as a histogram, which is then smoothed, a process shown schematically in Fig. 1b. The more k points used, the more "realistic" the computed $\varrho(e)$. The electronic energy of the system is then obtained by summation or integration as in Eqs. 4, 5.

for a molecule

$$E = 2\sum_{i=1}^{h} \varrho(e)e_i \tag{4}$$

for a solid

$$E = 2\int_{-\infty}^{e_F} \varrho(e)e\,de \tag{5}$$

where e_F is the Fermi energy, the energy of the highest occupied level in the solid, analogous to the HOMO (level h) of the molecule. In fact we are not usually interested in the total energies given by these equations. More often we are concerned with seeing which of two alternative structural possibilities is lower in energy. More specifically we are often keen to see how the answer depends upon electron count and thus place the responsibility for the computed (and hopefully observed) structural differences on one or perhaps two orbitals of the molecular orbital collection. Analysis of the atomic orbital composition of such levels then invariably tells us why the structural change occurs. In the solid the description of the orbitals around the Fermi level is of comparable importance. It is interesting then that much of the information contained in the diagrams at the far right-hand side of Fig. 1 is not used in understanding structural differences.

There is another method for computing densities of states which we will briefly outline here. It is based on the method of moments and, until recently, has been exclusively used by the physics community[3–14]. The moments of the electronic density of states are defined as in Eqs. 6, 7

for a molecule

$$\mu_n = \sum e_i^n \tag{6}$$

for a solid

$$\mu_n = \int_{-\infty}^{\infty} e^n \varrho(e) de \tag{7}$$

It is possible if we know the collection of moments $\{\mu_n\}$ to generate directly the electronic density of states $\varrho(e)$ (Fig. 2). Obviously for a molecule we are only interested in a small, finite number of the μ_n since we wish to generate a discrete spectrum. For the solid however, the larger the number of moments included in the inversion process, the better our computed $\varrho(e)$ mimics reality (Fig. 3 shows this for the pπ orbitals of graphite). In this sense recall our earlier comment concerning the traditional method of generating a $\varrho(e)$ diagram. However, there is one extremely important observation concerning both approaches which will be invaluable in our discussion later. Although to the eye the shape of the density of states plot is a slowly converging function of the number of moments used in its generation, the energy obtained via use of Eq. 5 converges quite quickly[10]. In other words if we are interested in energies, then a small number of k points

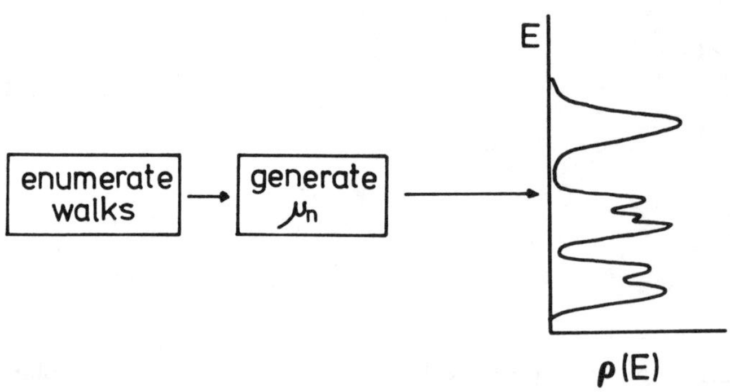

Fig. 2. Schematic showing the direct generation of the energy density of states by inversion of the $\{\mu_n\}$

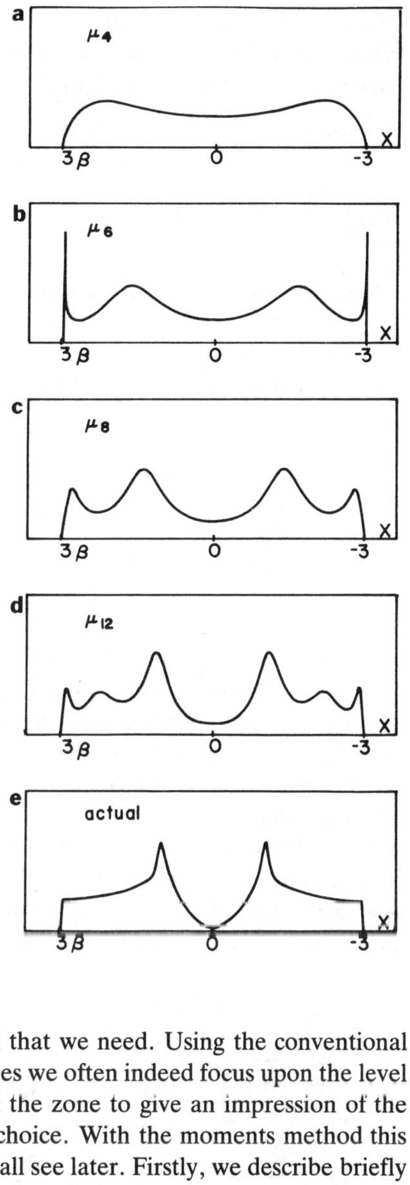

Fig. 3. Computed energy density of states for the pπ levels of graphite as a function of the number of moments used in the inversion process

or a small number of moments may often be all that we need. Using the conventional method of generating an electronic density of states we often indeed focus upon the level changes at Γ and other high symmetry points in the zone to give an impression of the electronic factors behind a particular structural choice. With the moments method this result has some far reaching implications as we shall see later. Firstly, we describe briefly the rather interesting way the moments themselves may be generated from the collection of atomic orbitals in the molecule or solid.

B. Generation of Moments

The n-th moment of the electronic density of states is given by either Eq. 6 or 7. It is very easy to show that within the confines of simple Hückel theory

$$\mu_n = \sum_i e_i^n = \mathrm{Tr}(\boldsymbol{H}^n) \tag{8}$$

where **H** is the Hamiltionian matrix of Eq. 1. H is Hermitian, so there is a unitary matrix S such that $S^{-1}HS$ is diagonal with elements that are the eigenvalues of H namely the e_i. Then $\mathrm{Tr}(H^n) = \mathrm{Tr}[(S^{-1}HS)^n] = \sum_i e_i^n$. Perhaps the most interesting aspect of this expression is seen when the Trace is expanded algebraically[3)] as

$$\mathrm{Tr}(H^n) = \sum_{i_1, i_2 \ldots i_n} \mathbf{H}_{i_1 i_2} \mathbf{H}_{i_2 i_3} \ldots \mathbf{H}_{i_n i_1} \tag{9}$$

1 shows geometrically a single term in this sum. It represents a path starting and ending

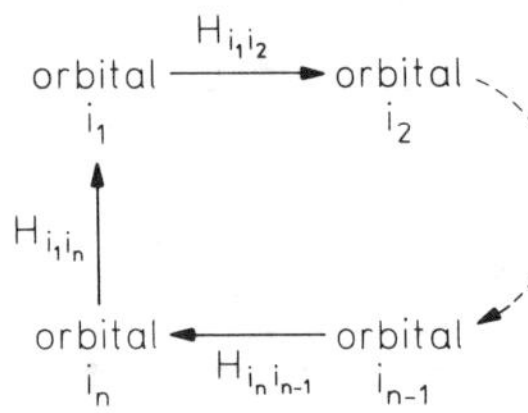

1

with orbital i_1. The weight associated with each step in the path is simply the $H_{i_a i_b}$ interaction integral (β in the Hückel scheme) which links the orbitals i_a, i_b. Thus the nth moment is simply the weighted sum of all paths which start off at orbital i_1 and return in n steps back to this orbital. If one of the steps in the path connects orbitals which do not interact with each other, then the weight of the whole path is identically zero. This geometrical interpretation of the nth moment will be invaluable to us later since it permits an energetic description of solids and molecules in terms of the connectivity of the orbitals (and hence atoms) of the system. Walks "in place" in the sum of Eq. 9 will lead to entries of the tpye $H_{i_a i_a}$, namely Hückel α contributions. Without any loss of generality we can in many cases set $\alpha = 0$ and remove such walks from consideration.

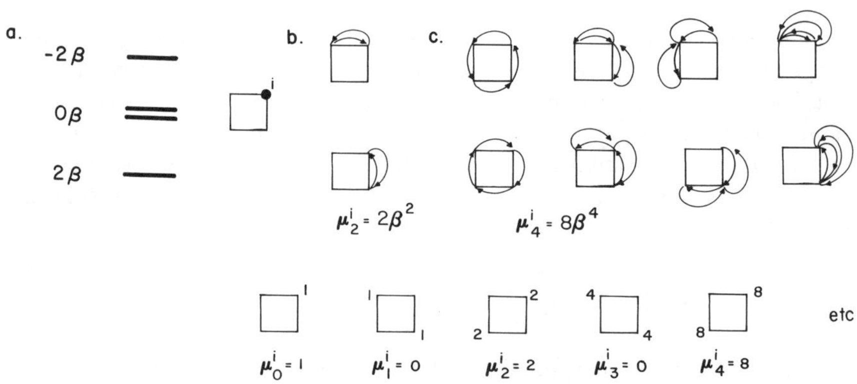

Fig. 4a–c. Enumeration of the moments of the pπ levels of cyclobutadiene. **a)** Shows the energy levels referred to $\alpha = 0$. **b)** enumerates the walks explicitly. **c)** Shows how the same result maybe obtained by propagating a unit vector around the ring

With this extra simplification the enumeration of the walks and generation of the moments for the pπ levels of cylcobutadiene proceeds as shown in Fig. 4. As we will see later walks "in place" will need to be incluced whenever there are atoms of different electronegativity present.

One of the interesting implications of this section is that the walks, and hence moments, may be generated without any recourse to the translational symmetry of the solid, or the point group of the molecule. In the context of extended arrays therefore, the moments method may be used in the study of aperiodic systems such as are found[3–9] in amorphous materials and in surface phenomena. In this article we shall exclude such areas fom discussion, and will concentrate on structural problems in molecules and crystalline solids.

We shall also use the lowest common denominator to derive the energies of our examples, namely the Hückel approach, with the proviso that the value of β will depend upon the nature and orientation of the interacting orbitals. In other words we shall write $\beta = \sum_{\lambda} \beta_\lambda S(\lambda, \theta, \varphi)$ where $\lambda = \sigma, \pi, \delta$ and (θ, φ) describes the orientation of the orbitals.

C. Energies as a Function of Electronic Configuration

The details of the inversion process shown schematically in Fig. 2 do not in fact concern us in this article. The actual methods used are described in several places[3–10, 14]. In Ref. 14 we show how the eigenvalue spectra for simple molecules and solids may be constructed, and in Ref. 15 compare the two methods shown in Figs. 1 and 2. We are particularly concerned though with the energy difference between two systems as a function of electronic configuration. Recall an important result from our earlier discussion. Although the energy density of states is often a slowly converging function of the number of moments included in the inversion process, the electronic energy usually converges rapidly. Let us consider then two systems I, II for which the first m moments are the same, because of their geometrical sturcture. Then use of these m moments for I and II to generate an energy difference curve will lead to a value of zero for all electron counts. Only on inclusion of the first moment which is different (m + 1) will a sensible energy difference curve result. Since the energy does converge rapidly with the number of moments included, consideration of just the first disparate moment or perhaps the first couple of disparate moments might be all that is needed to understand the variation in

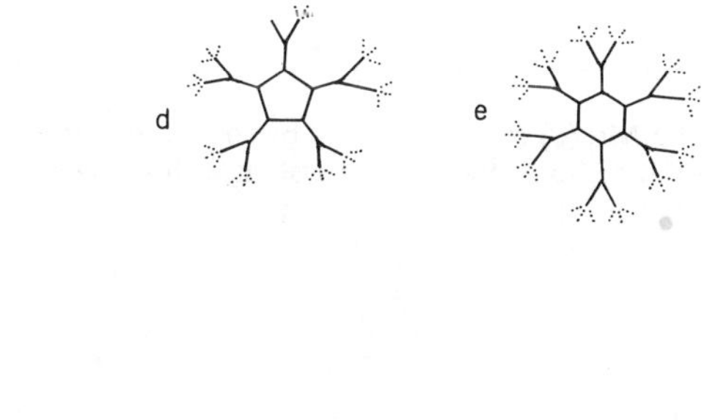

Fig. 5. a) A three tree. **b)–e)** Three trees containing rings

electronic stability with electron count. What we wish to derive then is the relationship between the energy difference curve as a function of electron count associated with two structures and the geometrical description of those systems in terms of loops of the type shown in **1.** This may be done in several ways but there is one approach[10] which is pictorially a useful one. Figure 5a shows a Cayley tree which proceeds indefinitely in two dimensions. Each node is three coordinate. Figure 5b shows a three membered ring in the middle of this tree. We can locate pπ orbitals at each node of both of these nets, calculate the energy density of states, and hence the energy difference curve between these two hypothetical structures as a function of electron count. The major difference between the two systems is that 5b differs from 5a at the third moment since there is an extra walk of length three in Fig. 5b which goes around the ring (**2**). By inclusions of rings of different sizes in the tree as shown in Fig. 5c–e we may similarly generate energy difference curves for systems which differ at the fourth, fifth and sixth moments respectively.

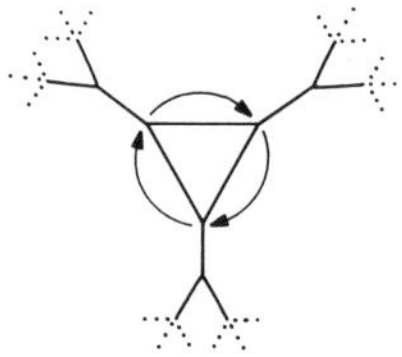

2

The energy difference curves that result, $\Delta E_r(x)$ are shown[10, 11] in Fig. 6. The ordinate x represents the fractional orbital occupany. At x = 1 all the orbitals of the problem are doubly occupied with electrons. Notice that the amplitude of these curves decreases

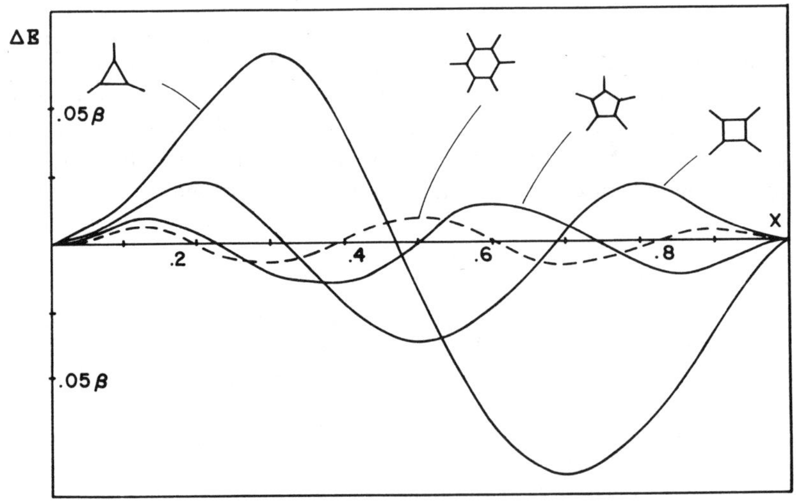

Fig. 6. The ΔE_r (x) curves obtained by comparing the energy density of states of the species in Fig. 5. When the curve is positive the system with the larger $|\mu_r|$ is more stable. (One of the curves is dashed for clarity)

as the ring size gets larger. This is in accord with our statement earlier that the electronic energy converges rapidly with the number of moments included in the inversion process. Notice too that the curves have characteristic shapes, the number of nodes being equal to the ring size if those at x = 0.1 are counted. The reader can clearly generate the form of the curves for higher moment differences.

The decrease in the energetic importance of the nth moment as n increases is of course quite understandable. First nearest neighbors are linked to the home atom by walks of length two (there and back) and of length three (via one nearest neighbor to another). Second nearest neighbor effects arise at the earliest in the fourth moment since this represents a walk in two steps to this further neighbor and back. Accordingly, third nearest neighbor effects will manifest themselves at the earliest in the sixth moment. Intuitively we expect the largest energetic effect to occur for the atoms directly bound to a given atom and the influence of neighbors further away to be less important, in fact dropping off in importance as the distance increases. Such a view is in accord with the moments results.

It is quite illuminating to use these results to comment[11] on the stability of cyclic polyenes containing rings of various sizes compared to their open-chain analogs. The first moment which will be different for the pπ orbitals of an m-ring compared with those of the open chain, or of a ring of large size, will be the m-th. Since in each neutral C_mH_m molecule there are m pπ orbitals and m electrons, the fractional orbital occupancy is 0.5.

Figure 7 shows where various planar carbon rings lie on the curve of Fig. 6. Notice that the results encompass Hückel's 4n + 2 rule[1] quite nicely. The electron counts predicted by this rule to give stable structures lie in regions where the m ring is stabilized with respect to the molecule which contains no such loops. Thus we find stability for C_6H_6 but not $C_6H_6^{\pm 2}$, stability for $C_4H_4^{\pm 2}$ but not C_4H_4 itself, etc. The moments method has therefore quickly provided a new way of looking at a very old problem.

Although the energy difference curves of Fig. 6 have been generated by using a model based upon the overlap of pπ orbitals, similar types of curves will arise for other orbital problems involving mth moment differences. Obviously the energy scale of the plots will be very sensitive to the nature of the atoms involved and to the type of interaction between the relevant orbitals i.e., whether it is of σ, π or δ type. Also the location of the crossing points will change a bit from system to system, but the general features of Fig. 6 will remain. In the rest of this article we will examine a series of structural problems, which at first sight appear to be unrelated but which may be put on a common footing by using the results we have just described.

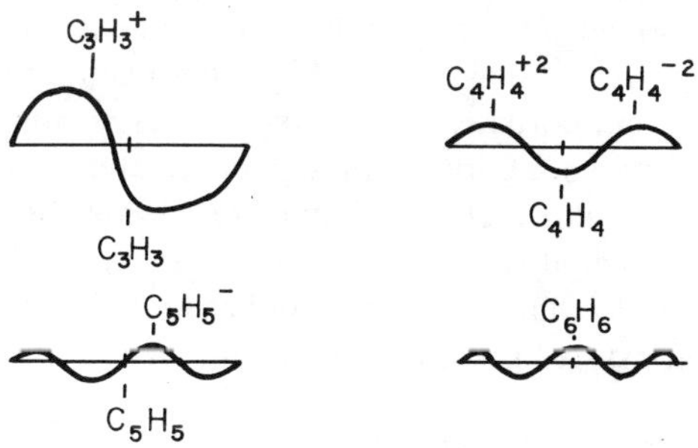

Fig. 7. Superposition of the electron counts appropriate for the pπ levels of small rings on the curves of Fig. 6

III. Some Structural Problems

A. Graphite and Related Nets

Structures **3–6** show the graphite net and some alternative arrangements which contain rings of different sizes. Clearly the first moment which is different when compared to graphite is the third, fourth and fifth for **6, 4** and **5** respectively since these structures allow walks around rings to these sizes not available in the graphite structure. Figure 8a

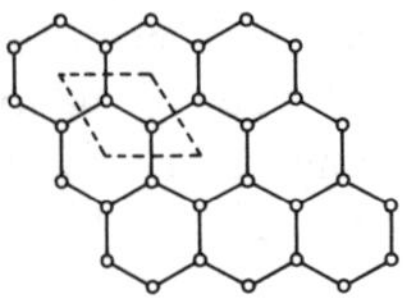

3

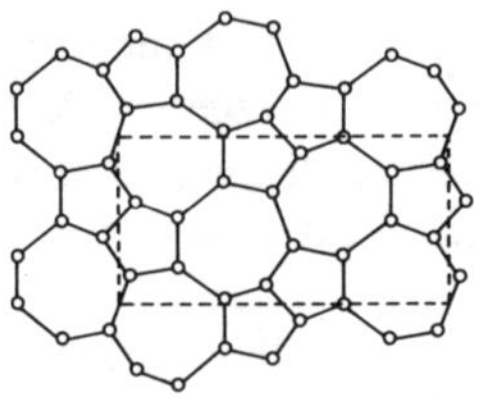

5

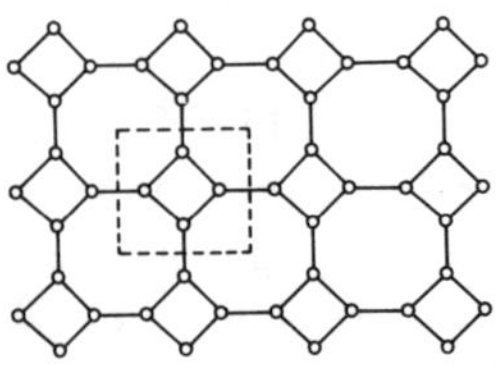

4

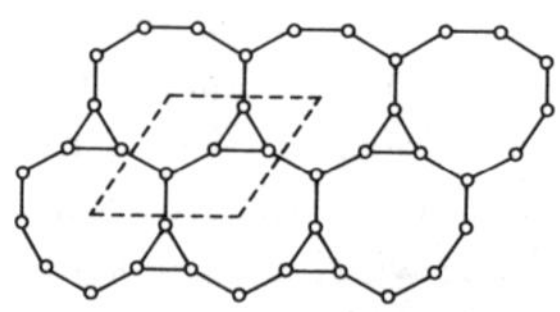

6

shows the computed[11] energy differences between **3** and **4, 5** and **6** as a function of electron filling. Notice **4** is more stable early on, but just before the half filled point becomes less stable than graphite. The shape of this energy difference curve is typical (Fig. 6) of two systems which differ at the third moment. Structure **5** is less stable than graphite at the half filled point but becomes more stable at the quarter and three-quarter filled positions. This is just what is expected for two systems which differ at the fourth moment. Similarly the energy difference curve associated with graphite and **5** has the features associated with the curve expected for two systems which differ at the fifth moment. However notice that while the n = 5 curve from Fig. 6 has a node at x = 0.5 the corresponding node in Fig. 8 has been shifted to higher x. This has occurred as a result of the importance of the six-membered rings in graphite itself. In other words the energy difference curve, although dominated by the fifth moment, also has a contribution from the sixth moment. Figure 8b shows the curves that result by constructing a suitably weighted (by the numbers of rings) linear combination of the $\Delta E_r(x)$ curves of Fig. 6. Notice that the agreement between the two halves of Fig. 8 is quite good and implies that for systems of this type we may write a simple expression for the energy difference curve.

$$\Delta E(x) = \sum \Delta N_r \Delta E_r(x) \tag{10}$$

where ΔN_r is the difference in the number of rings of size r in the two structures.

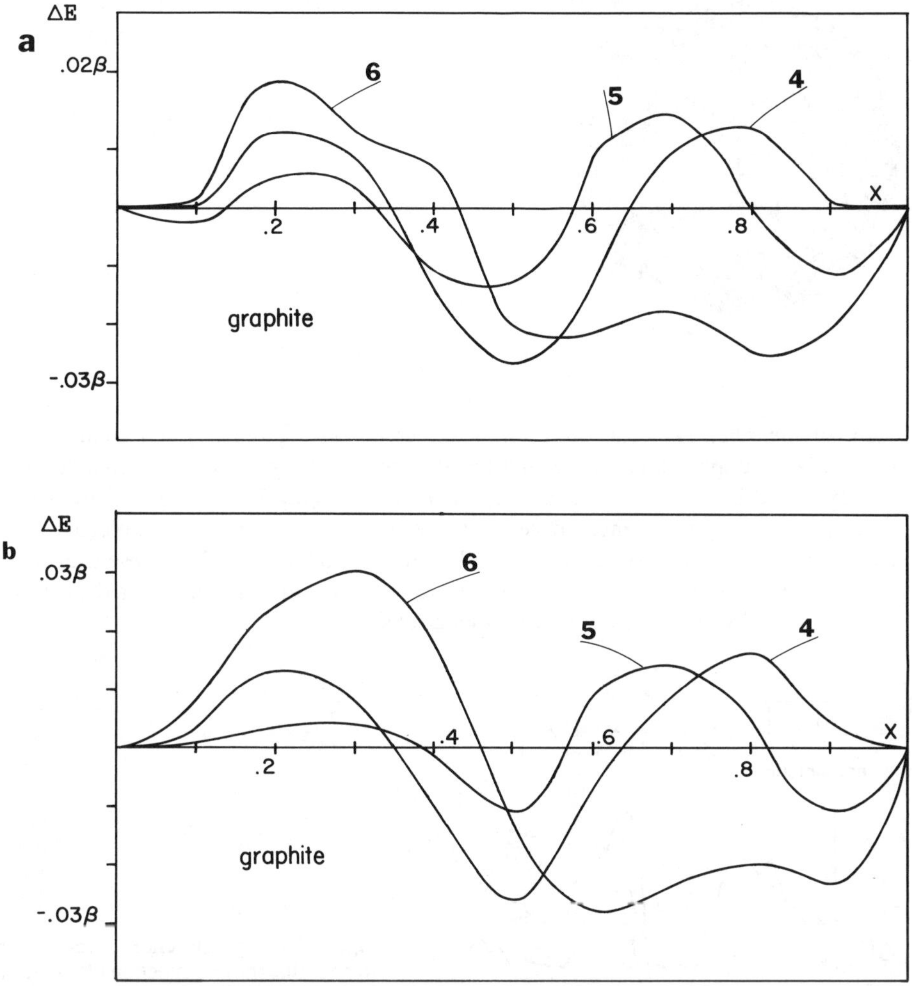

Fig. 8 a, b. Energy difference curves between graphite and the other structures **4–6. a)** Computed difference curve. **b)** Simulated curve obtained by adding the densities of states of the corresponding units of Fig. 5, i.e. expressing the energy differences as a linear combination of the $\Delta E_r(x)$ of Fig. 6

Of interest is the fact that whereas graphite is indeed the structure observed at the half-filled point (and computed to be the geometry of lowest energy) the net containing 5 and 7 membered rings is favored from the calculations for somewhat higher electron counts. It is in fact found for the nonmetal net in ScB_2C_2. Assuming the metal is present as Sc(III) then x = 0.625 for this system in nice agreement with the computed energy difference curves of Fig. 8a. Could we induce such a structural change by intercalation of graphite by an electron donor, such as an alkali metal? Probably not in fact since several CC linkages will need to be broken and remade for this to happen. Another intriguing possibility is shown in **7.** It can be shown that a hexagonal net, no matter how big, can be folded into a polyhedron by inclusion of twelve pentagons. **7** shows in fact an enlarge-

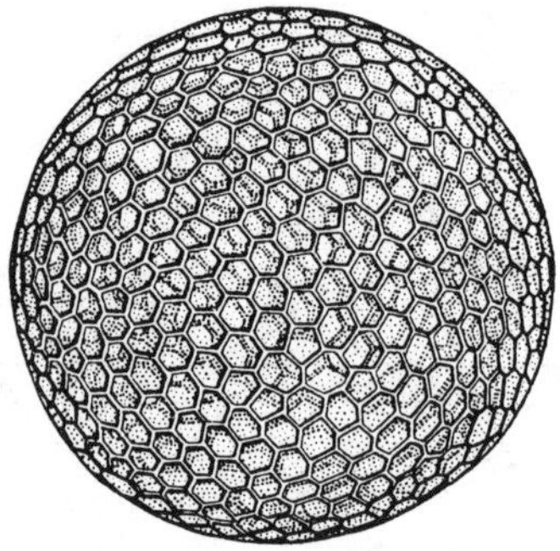

7

ment of the silica skeleton of *Aulonia hexagona,* one of the tiny sea creatures called radiolaria. A dopant level somewhat higher than currently used in semiconductor technology would be required to produce a sphere containing 260,000 carbon atoms[16].

Similar energy difference curves to those of Fig. 8 are found for molecular species containing these four, five and six-membered ring building blocks. These are shown in Fig. 9. The organic chemist is only interested of course in the result for the all carbon system (x = 0.5) with a total of 10 pπ electrons.

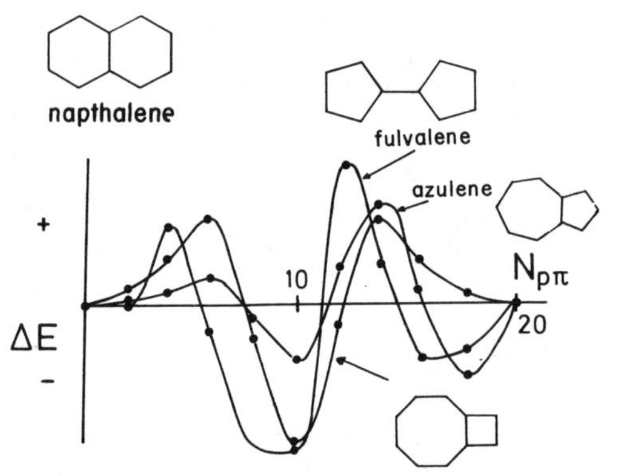

Fig. 9. Energy difference curve compared to naphthalene as a function of pπ electron count for the three organic molecules shown. Notice the similarity in shape to the curves of Fig. 8

B. The Crystal Structures of Tl, Pb, Bi and Po

The crystal structures of the elements are given in Table 1. Tl and Pb crystallize in one of the two simplest close-packed arrangements, hcp and fcc respectively. These structures (Fig. 10), built up from layers of close-packed planes, contain a multitude of three membered rings. The structure of α-Bi (Fig. 11) may be regarded as a puckered analog of graphite. Here there are only six-membered rings. The structure of α-Po (Fig. 12) is simple cubic and contains four membered rings. We then expect an energy difference curve for α-Bi and α-Po to be dominated by the fourth moment, and that for α-Bi and the fcc and hcp structures to be dominated by the third moment. For these heavy elements at

Table 1. Crystal structure and other data for the elements Tl–Po

	Tl	Pb	Bi	Po
configuration	s^2p^1	s^2p^2	s^2p^3	s^2p^4
x (p band)	0.167	0.33	0.50	0.67
structure	hcp	fcc	α-Bi	simple cubic
dominant ring	3	3	6	4

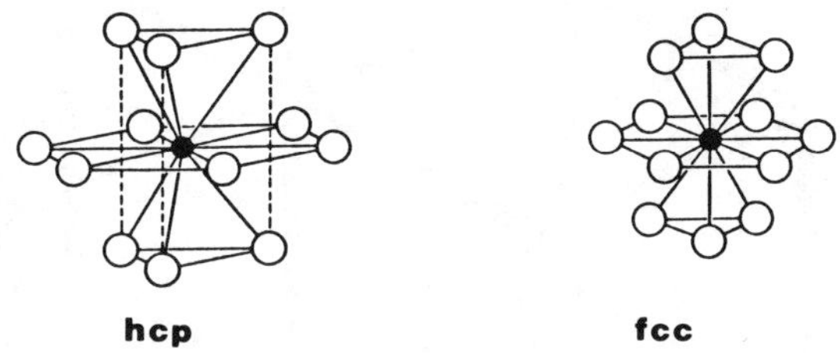

Fig. 10. The local coordination in hexagonal and cubic close packed arrangements

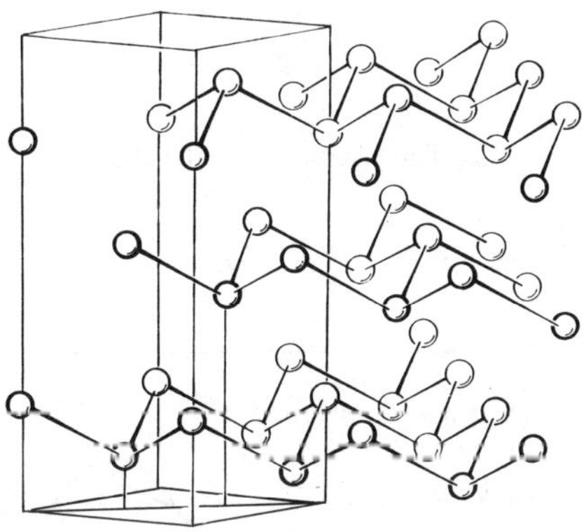

Fig. 11. The α-Bismuth (or α Arsenic) structure, a puckered analog of graphite

the bottom right-hand side of the periodic table the $6s^2$ pair of valence electrons often behaves in a core-like fashion, the so-called inert pair. As a simplification we shall ignore them. Table 1 then shows the values of x for an orbital problem associated purely with the 6p orbitals of the elements. Figure 13 shows a trio of computed[11)] energy difference curves using this p orbital only model. Notice how the expectations of the moments method are nicely reproduced. The close-packed structures are stable at early band fillings, p^1 and p^2, α-Bi with its six-membered rings is stable at the half-filled p band (p^3) and α-Po is calculated to be stable (just) at the two-thirds full band (p^4).

The fcc and hcp structures differ at the second nearest neighbor level and so their first disparate moment will in principle be the fourth. In fact their fourth moments are very similar and it is the fifth moment which is different. **8** shows the fifth moment energy difference curve which enables resolution of these two structures. It predicts that Tl should adopt the hcp and Pb the fcc structure as indeed observed. The different walks of length five which give rise to this effect are shown pictorially in Ref. 11. They are

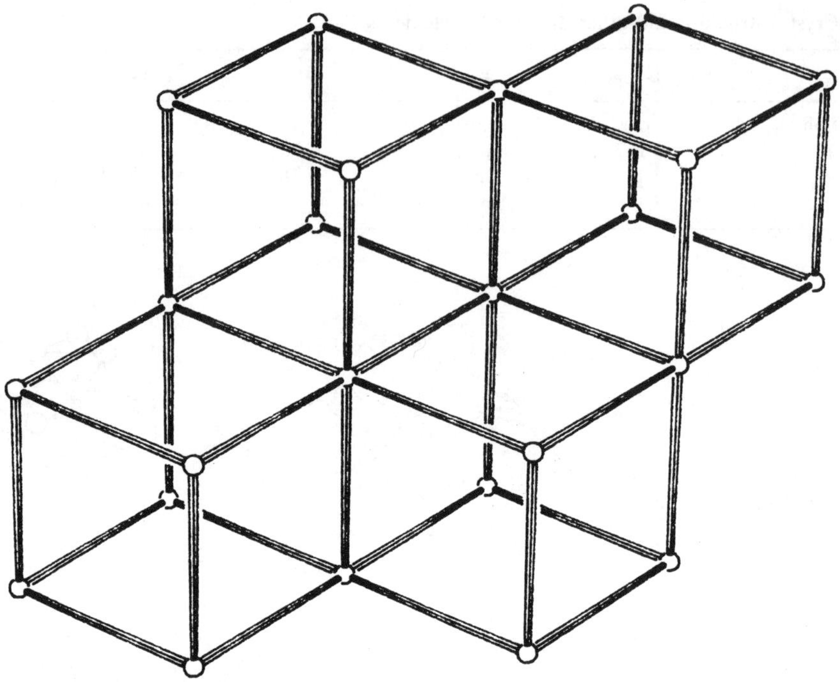

Fig. 12. The simple cubic structure of α-Polonium

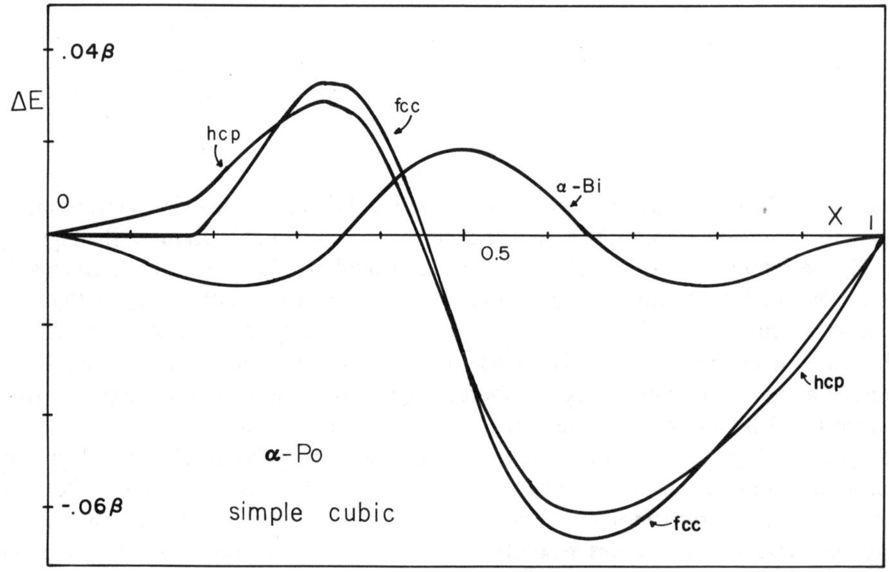

Fig. 13. Computed energy difference curves for the structures of the heavy elements of Figs. 10–12

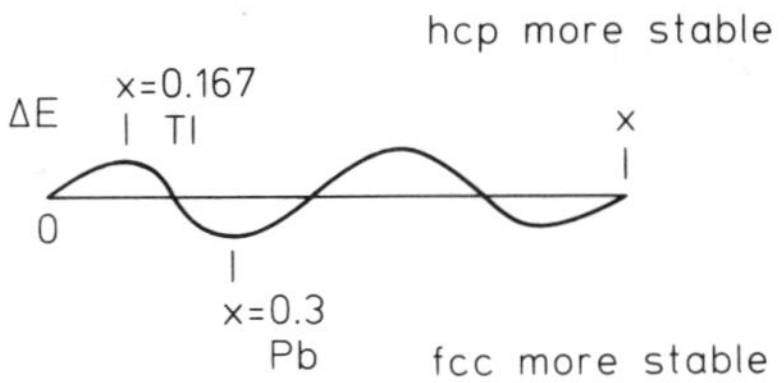

8

complex enough to prevent us from making a simple statement, based on geometry which will immediately identify the feature which electronically distinguishes them. Here then is a restriction of the method, sometimes it does not help us too much.

C. The Geometry of Allene

The D_{2d} geometry of allene is shown in **9** and is the one expected on the basis of van't Hoff's famous rules of stereochemistry. The D_{2h} geometry, the anti van't Hoff arrangement is shown in **10.** Why is the D_{2d} geometry preferred? Figure 14 shows Hückel

twisted

9

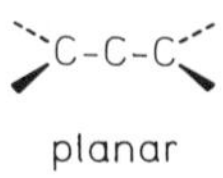

planar

10

molecular orbital diagrams for the π orbitals of the two arrangements and Table 2 the stabilization energy as a function of the number of pπ electrons. Clearly the results of the Hückel calculation are in accord with the observed allene geometry. From the point of view of the moments approach we are interested in the walks of various lengths which are different for the two structures. Both arrangements have walks of length 4 which are of the type shown in **11** but the D_{2h} geometry has extra walks (four of them in fact) of the type shown in **12.** So the energy difference curve as a function of the number of pπ

11

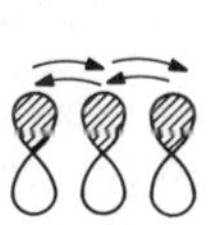

12

Table 2. Stabilization energies of planar and twisted allenes

Number of π electrons	Planar	Twisted	Difference
0	0	0	0
2	2.828 β	2 β	0.828 β
4	2.828 β	4 β	− 1.172 β
6	2.828 β	2 β	0.828 β
8	0	0	0

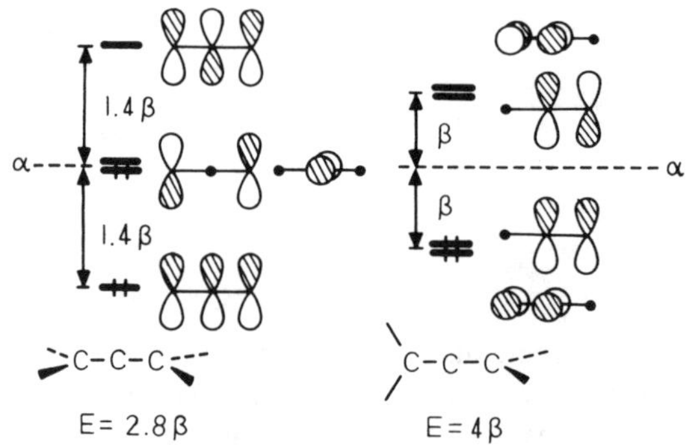

Fig. 14. Hückel energy level diagrams for the two allene geometries of **9** and **10**

electrons should resemble the fourth moment curve of Fig. 6. There will be a small difference produced by the discrete nature of the problem in that the plot will be made up from a series of points. Structure **13** shows the result obtained using the figures of

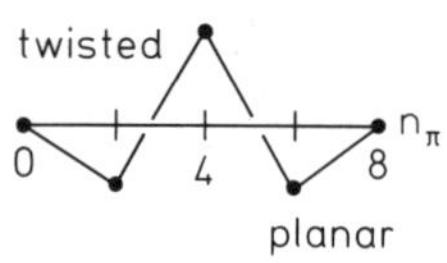

13

Table 2. It does indeed have a characteristic fourth moment shape and predicts that anti van't Hoff stereochemistry should be possible at electron counts other than the one for allene itself. Calculations[17)] by Schleyer and coworkers have suggested that this may well be the case. Structure **14** shows an allene derivative which contains an electron deficient three membered BBC ring. With two electrons assigned to a BB π bond the $(B_2H_2)C$ unit

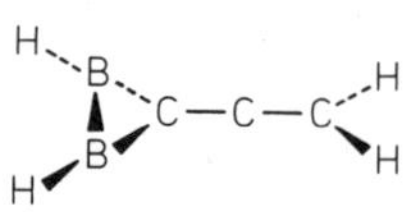

14

is then two electrons short compared to $(H_2)C$. The molecule **14** now contains only 2 π electrons in its allene π skeleton to be compared with the 4 π electrons of allene itself. The planar geometry is now the preferred one.

The van't Hoff rule applies to inorganic molecules too. For example the $Si_2NBeNSi_2$ skeleton in $\{[(CH_3)_3Si]_2N\}_2Be$ is arranged in the same fashion as that in the isoelectronic species allene itself.

D. Linear and Bent AH_2 Molecules

For simplicity let us use the Rundle-Pimentel scheme[18)] for this problem. One pair of electrons is stored in the central atom s orbital and the angular geometry is controlled by the interaction of p orbitals on A with the hydrogen 1 s orbitals. Figure 15 shows molecular orbital diagrams for linear and bent geometries constructed using the angular overlap model[18]. The energy differences between the two geometries are shown in Table 3. Notice how the predicted geometry varies with electron count in agreement with experiment. (The electron count included for "NeH_2" is 6, OH_2 is included with CH_2 since the p_z orbital shown on the diagram can accommodate a (lone) pair of electrons). It is a simple matter to enumerate the walks of different lengths associated with these two geometries. We are of course only interested in extra walks that arise in one arrangement compared to another. **15** shows the type of extra four-walk that is possible in the linear

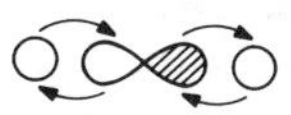

15

geometry but not in the bent one. (There are in this configuration four walks of this type). So the linear geometry has the higher fourth moment and so is the geometry which is not favored at the half-filled position in agreement with experiment (see Table 3). **16** shows the energy differences of Table 3 in pictorial form and the fourth moment nature of this problem is now apparent. Notice that the stabilization of the linear geometry at the quarter and three quarter positions is dominated by the second order, e_σ, term but the stabilization of the bent geometry at the half-full point by the quartic, f_σ, term. This plot of course is immediately comparable to the allene example of the previous section where all of the orbitals of the problem were identical. Notice, however, that the

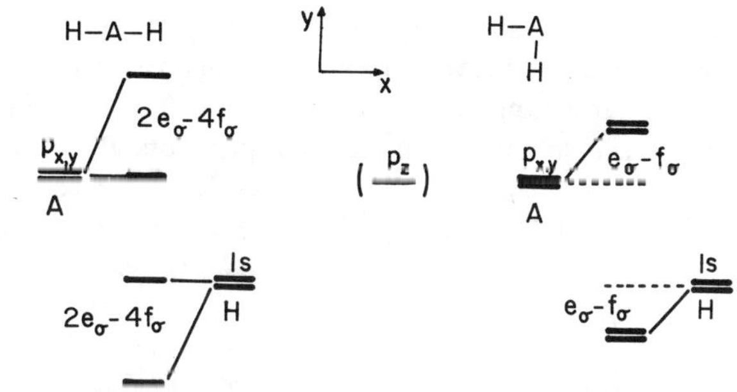

Fig. 15. Molecular orbital diagrams for linear and bent AH_2 molecules, using a p-orbital only model and constructed using the angular overlap approach

Table 3. Stabilization energy of linear and bent AH_2 molecules as a function of electron count

Number of electrons[a]	Linear	Examples	Bent	Difference
0	0		0	0
2	$4e_\sigma - 8f_\sigma$	BeH_2	$2e_\sigma - 2f_\sigma$	$2e_\sigma - 6f_\sigma$
4	$4e_\sigma - 8f_\sigma$	CH_2, OH_2	$4e_\sigma - 4f_\sigma$	$-4f_\sigma$
6	$4e_\sigma - 8f_\sigma$	(NeH_2)	$2e_\sigma - 2f_\sigma$	$2e_\sigma - 6f_\sigma$
8	0		0	0

[a] In p_x,p_y orbitals only

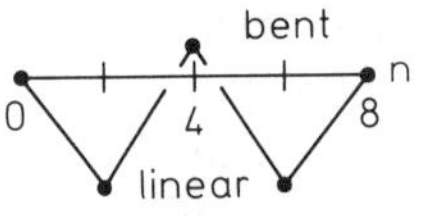

16

amplitude of the plot at the half-filled point is now much smaller than that at the quarter and three-quarter points, in contrast to the results for allene. The problem of linear and bent triatomics will be discussed again later in this review.

E. Cis and Trans Transition Metal Dioxo Compounds

In orbital terms this is a small variation on the AH_2 problem above. d^0 transition metal dioxo compounds are always found in the *cis* arrangement, while d^2 compounds, of which there are a small number of examples[19)], are found in the *trans* arrangement. For example d^2 $[Mo(CN)_4O_2]^{4-}$ has the trans structure but d^0 $[Mo(CN)_4O_2]^{2-}$ the cis arrangement. The d^2 ion $MoO_2Cl_4^{4-}$ also has the two oxo ligands arranged in a *trans* fashion. (There are of course many actinide dioxo systems which contain a *trans* MO_2 unit. We shall exclude these from discussion). Figure 16 shows two molecular orbital diagrams, constructed for *cis* and *trans* dioxides using the angular overlap model. The problem has been pared down to its essentials for simplicity and we show just the interactions involving the π manifold of orbitals and dominated by the π donor characteristics of the oxide ligand. Notice that such a model correctly predicts the changeover in preferred structure with electron count. Again this is a fourth moment problem. In the linear, *trans,* geometry the two sets of oxygen pπ orbitals share two central atom orbitals (xz,yz). In the bent, *cis,* geometry they share only one. In the latter therefore the number of four walks which link one ligand orbital with another through the central atom d orbitals is half as large. It is the linear arrangement then which has the larger fourth moment and therefore the one which is disfavored at the d^0 configuration. Just as in the AH_2 case the bent geometry is favored at d^2 by energetic terms which arise in second order (e_σ) but the linear geometry is favored at d^0 by terms which arise in fourth order (f_σ). Importantly however notice how

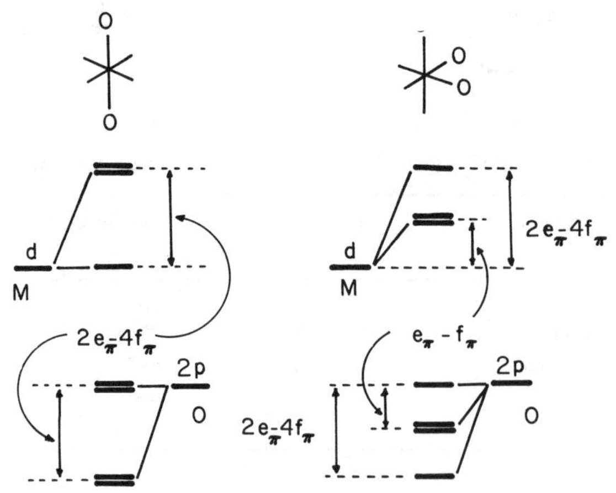

Fig. 16. Molecular orbital diagrams for cis and trans MO_2L_4 species, emphasizing the (p-d)π interactions between the metal and oxygen atoms

the crossing points of this curve have moved compared to those found for $\Delta E_4(x)$ of Fig. 6.

F. The Rutile and α-PbO_2 Structures

Figure 17 shows the structures of rutile and α-PbO_2, two structures based on an hcp net of oxide ions in which half of the octahedral holes are filled by cations. In rutile straight

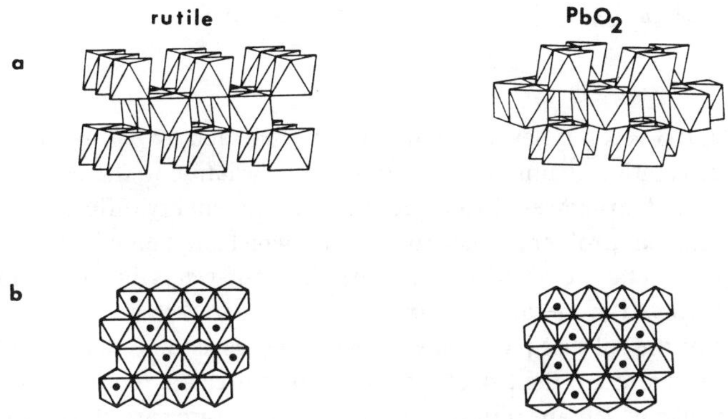

Fig. 17 a, b. The rutile and α-PbO_2 structures showing **a)** the straight and zigzag chains of MO_6 octahedra and **b)** the way the interstices in the hcp oxygen arrays are filled by metal ions. (Both structures are actually distorted away from these idealized pictures)

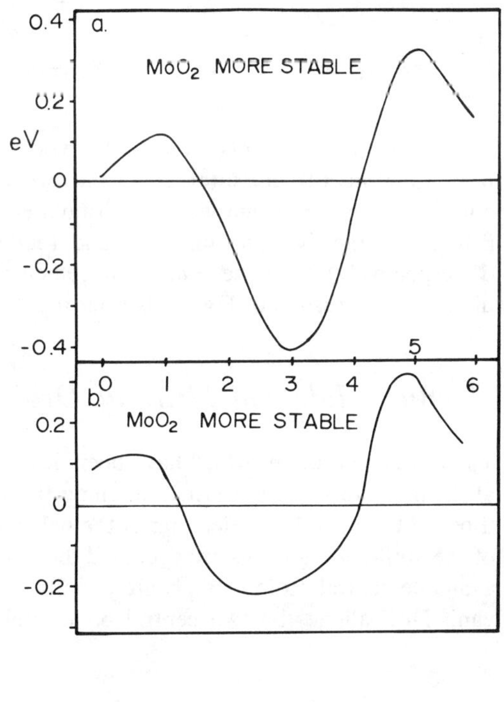

Fig. 18 a, b. Computed energy difference curve between rutile and α-PbO_2 structures as a function of d count **a)** Curve for rutile and α-PbO_2 and **b)** Curve for the distorted variants MoO_2 and β-ReO_2 as shown in **17**

chains of *trans* edge-sharing octahedra are produced. In the α-PbO_2 structure the chains are zigzag ones since skew edges of the octahedra are shared. For electron counts between d^0 and d^6 pairing up of the metal atoms in these structures are often found. The rutile type is simply related to the MoO_2 type by such pairing (shown in **17**). The α-PbO_2

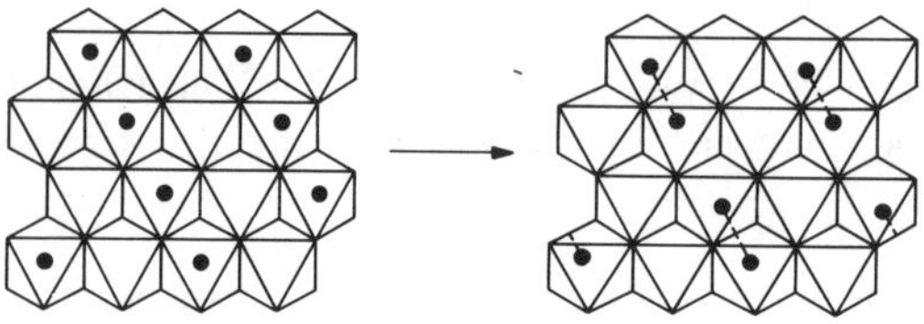

17

and β-ReO_2 types are analogously connected by pairing. Figure 18 shows the results of a calculation[20] aimed at evaluating the variation in the energy difference between the two sets of structures. From the shape of the energy difference curve this is clearly a fourth moment problem. Since the orbitals which are being filled in Fig. 18 are the t_{2g} orbitals on the metal we need to examine the four-walks involving these orbitals which will be involved in metal-metal interactions. There are a pair of walks which are critical ones. For the sake of pictorial simplicitiy, let us assume that the edges of the octahedra which are shared in the β-ReO_2/α-PbO_2 structures are the *cis*, rather than skew ones. Then the walks which are different in the two structures are those which are shown in **18** and **19.** The walk of **18** is weighted by β_π^4 but the corresponding one in **19** is only weighted by

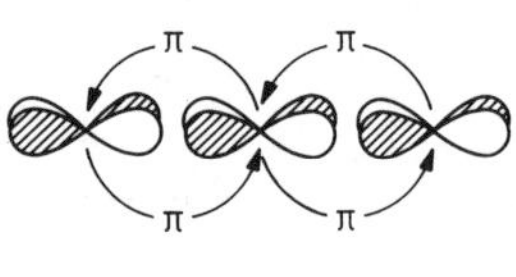

18

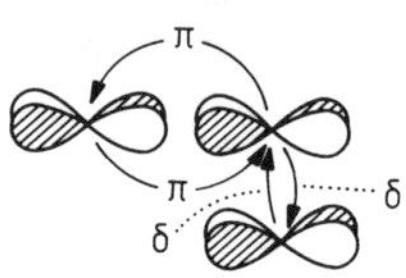

19

$\beta_\pi^2\beta_\delta^2$ where the subscripts indicate the type of orbital overlap involved. Since δ interactions are weaker than π interactions the *trans* arrangement has the higher fourth moment and therefore is the one which is disfavored at the half-filled point (d^3). Interestingly β-ReO_2 has the zigzag arrangement and it indeed is a d^3 system. MoO_2 and OsO_2 (d^2 and d^4 respectively) have the *trans* arrangement[20]. So, although the qualitative trend predicted by the results of Fig. 18 is fine, the details are not as good.

G. Busy, Idle and Metriotic Orbitals

The last few problems which have been described in terms of fourth moment differences all fit into the same pattern concerning the utilization of the central atom orbitals by those of the ligands. Notice that at the half-filled point, where all the bonding and none of the antibonding levels are occupied the most stable arrangement is the one where the available central atom orbitals are equally shared by the ligand orbitals. In the case of van't Hoff allene the two central $p\pi$ orbitals were both equally involved with the π

orbitals of the ligands. For the anti van't Hoff structure one central atom pπ orbital could interact with both ligand pπ orbitals while the other could interact with neither. In other words these half-filled problems are most stable in a geometry which avoids the generation of **busy** (over utilized) or **idle** (under utilized) orbitals and is energetically best where as many central atom orbitals are as equally used as possible. We call these orbitals **metriotic** from the Greek μετϱιτης meaning "the right amount or degree of anything". We have emphasized this point elsewhere[18, 21a)] in connection, not only with the allene geometry, but also in connection with the geometries of both main group and transition metal complexes. In terms of moments it is those walks of the type shown in **12** and **15** which allow one ligand orbital to "see" another via a central atom orbital which destabilize the busy orbital geometry at the half filled point. For other electron counts however, the busy/idle orbital combination may well be stable as indicated both by Schleyer's electron deficient allenes[17)] and the transition metal oxo examples. **20** and **21**

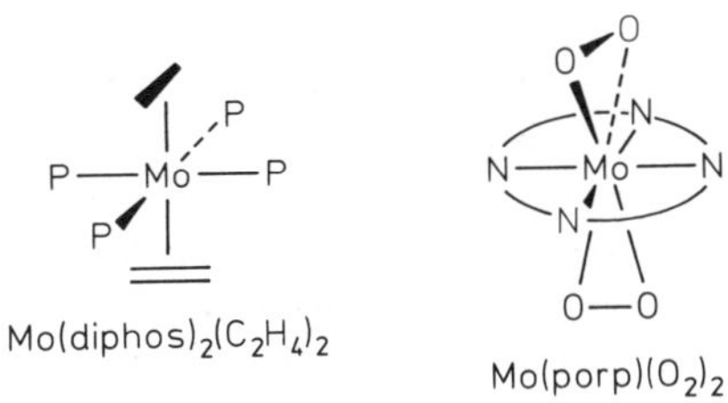

20

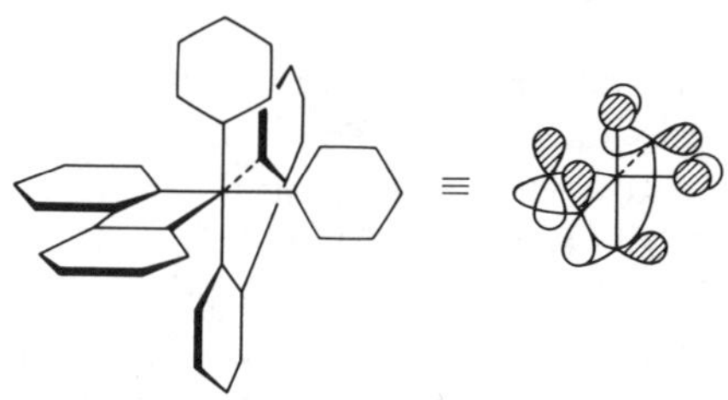

21

show some other examples[21)] of the same problem. The single-faced ethylene π-acceptor ligands in [Mo(diphos)$_2$(C$_2$H$_4$)$_2$] arrange themselves in a staggered way. Here the metal-ligand π bonding levels are full (the "t$_{2g}$" set of the d^6 metal) but the corresponding antibonding levels (ethylene located) are empty. In this light it is also interesting to note the stability of the cis or fac isomers for eighteen electron d^6 systems such as [M(CO)$_2$(diphos)$_2$] (M = Cr, Mo, W), [Mn(CO)$_3$(diphos)X], [Mo(CO)$_3$(PR$_3$)$_3$] etc. but the stability of the corresponding trans or mer isomers for their singly oxidized analogs. In this case the d^5 system has less than a half-filled complement of ligand π^* plus metal t$_{2g}$ orbitals, to be compared with the greater than half-full situation (of ligand π plus metal t$_{2g}$ orbitals) for the dioxo examples. The axial phenyl groups in [Cr^{III}Ph$_2$(bipy)$_2$]$^+$ are similarly (**21**) staggered with respect to one another. Another pair of examples which are slightly more complex, are shown in **22.** A simple model with which to describe this

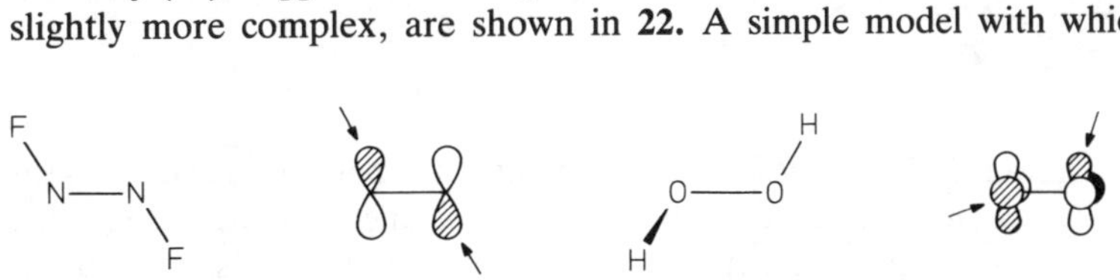

22

problem is shown in Fig. 19. It shows interaction of the two H atom orbitals with one or both components of the π^* levels of the A$_2$ molecule. The situation is identical to the one corresponding to the AH$_2$ molecules described earlier. This set of levels contains four electrons in the case of O$_2$H$_2$, but only two electrons for N$_2$H$_2$. Hydrogen peroxide, a

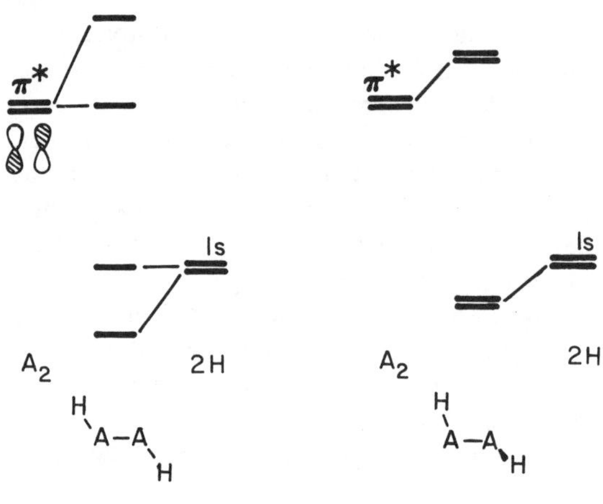

Fig. 19. Molecular orbital diagram showing interaction of two hydrogen ls orbitals with the π^* levels of O_2 in the O_2H_2 and N_2H_2 geometries. The angular overlap model is used

half-full problem, thus corresponds to a case where busy and idle orbitals are energetically infavorable. The converse applies to the quarter-full N_2H_2 case.

There are many transition metal containing examples which may be phrased in similar terms. Reference 21 contains a discussion of the preferred orientation of *trans* carbene ligands. It is also noteworthy that in d^6 transition metal complexes which contain CO, a good π acceptor ligand, the CO groups arrange themselves to be either *cis* or *mer* to each other rather than as the *trans* or *fac* alternatives.

The importance of metriotic orbitals is seen too in the structures of Jahn-Teller distorted solids[22]. **23** shows the cooperative effect in the structure of $KCuF_3$. In the Jahn-

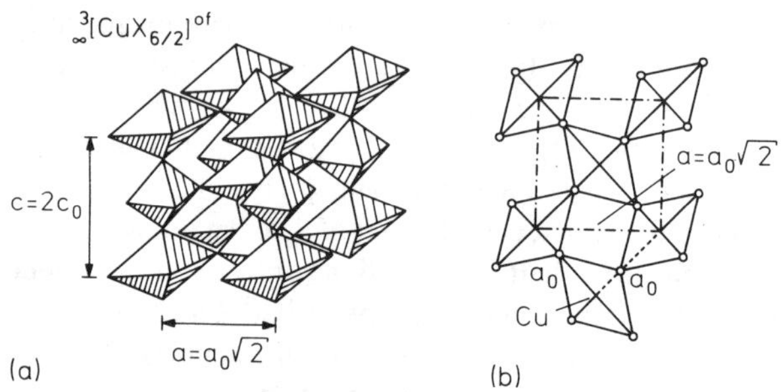

23

Teller inactive system $KZnF_3$ the perovskite arrangement is an undistorted one but in the Jahn-Teller active system containing Cu^{II} the distorted octahedra are arranged in a pattern termed antiferrodistortive. Here there are three different Cu–F distances, d_l, $d_m \approx d_s$ of 2.25 Å, 1.96 Å and 1.89 Å respectively. The d_m linkages are arranged exclusively parallel to the c axis but the d_l and d_s linkages are so arranged in (001) type planes such that they alternate along strings of copper atoms. So within these planes each fluorine atom is coordinated by one short and one long distance to copper (the metriotic situation) rather than the alternative geometry where half of the fluorines are coordinated by two d_s linkages (busy) and half by two d_l linkages (idle). The antiferrodistortive arrange-

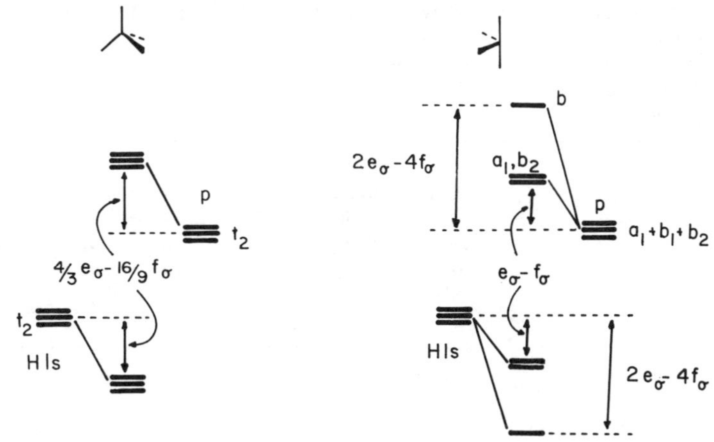

Fig. 20. Energy levels for tetrahedral and butterfly (SF_4) AH_4 molecules. A p orbital only model and the angular overlap approach are used

ment is the most common one and is always the one found when the octahedra share vertices as in the perovskite example of **23.**

Figure 20 shows p orbital only molecular orbital diagrams for tetrahedral and SF_4-like AH_4 molecules, constructed using the angular overlap model. Table 4 shows the energy difference between the two structures as a function of electron count and **24** shows the result pictorially. Notice again that this is a fourth moment problem as we mentioned above. Enumeration of the four walks and their weights is a little tedious for the tetrahedral geometry, but we can readily show that it is this structure which has the smaller fourth moment and therefore is the one which is more stable at the half-filled point, i.e., CH_4. Introduction of more electrons leads to an energetic preference for the SF_4 type of geometry. Notice from Table 4 and **24** that it is purely the difference in the quartic terms

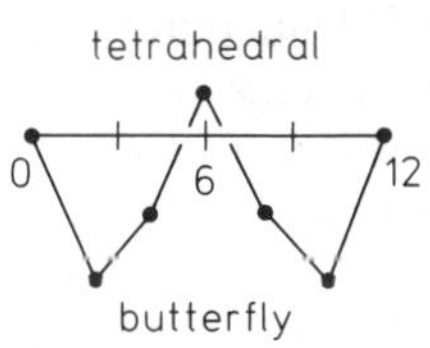

24

(f_σ) which leads to this result from the angular overlap model. There is a simple algebraical proof of the orbital sharing concept described above in terms of these f_σ parameters in Ref. 18. The smallest f_σ contribution will result when the three p orbitals all have the same energy, a result only possible for the tetrahedral geometry. In this geometry too, by symmetry all the p orbitals are equally utilized by the ligand orbitals, i.e., are metriotic. There are clearly three ways (at least) of stating the same orbital problem.

A closely related problem is that of the usage of the bond valence sum rule in structural chemistry. The bond valence may be defined as $s = A.\log(r/r_0)$ or $s = B(r/r_0)^N$ where A, N and r_0 are parameters for the given pair of atoms forming the bond (length r)

Table 4. Stabilization energy of tetrahedral and butterfly AH_4 molecules as a function of electron count

Number of p electrons	Tetrahedral	Butterfly [a]	Difference
0	0	0	0
2	$8/3e_\sigma - 32/gf_\sigma$	$4e_\sigma - 8f_\sigma$	$4/3e_\sigma - 40/9f_\sigma$
4	$16/3e_\sigma - 64/9f_\sigma$	$6e_\sigma - 10f_\sigma$	$2/3e_\sigma - 26/9f_\sigma$
6	$24/3e_\sigma - 96/9f_\sigma$	$8e_\sigma - 12f_\sigma$	$- 48/9f_\sigma$
8	$16/33_\sigma - 64/9f_\sigma$	$6e_\sigma - 10f_\sigma$	$2/3e_\sigma - 26/9f_\sigma$
10	$8/3e_\sigma - 32/9f_\sigma$	$4e_\sigma - 8f_\sigma$	$4/3e_\sigma - 40/9f_\sigma$
12	0	0	0

[a] SF_4 geometry

under consideration. In the absence of obvious anion-anion or cation-cation interactions (attractive or repulsive) the sum of the bond valences s around a center is remarkably invariant to its environment, and by definition may be set equal to its characteristic "valence". A secondary observation[23)] is that the set of bond valences characterizing the different linkages associated with a given atom are as equal as possible given the articulation of the structure. This is just the metriotic result expected.

H. Boron and Carbon Structures

An interesting feature of the structures found for the elements of Groups 13 and 14 is the prevalence of three membered rings in the former and their absence in the latter. Figure 21 shows the structure of cubic diamond and a hypothetical arrangement we call "tetrahedral" diamond. In this structure tetrahedra of atoms are located at the nodes of the cubic diamond structure. In both arrangements each atom is four coordinate. Figure 22 shows a computed energy difference curve[11)] as a function of electron count for the two structures. It has the appearance expected for a third moment problem. Indeed the tetrahedral diamond structure is full of three-membered rings but the cubic diamond structure has no ring smaller than six. Satisfyingly the cubic diamond arrangement is calculated to be more stable than the tetrahedral diamond arrangement at the half-filled point, the electron count corresponding to carbon.

Shown in Fig. 21c is the structure of another hypothetical diamond structure obtained by locating cubes at the nodes of a bcc lattice. Each atom is four coordinate here too. Figure 22 shows the energy difference curve[11)] for supercubane and cubic diamond, this time dominated by the fourth moment difference as expected from the squares present in the former. Notice that although the crossing points and the shapes of these curves are a little different from those of Fig. 6, the general features are quite similar.

Figure 23 shows the structure of rhombohedral boron, which may be considered as a close-packed array of B_{12} icosahedra. These deltahedra have a larger number of three membered rings. In comparing this structure energetically with that of carbon a problem arises. The coordination number in the boron arrangement is six but four in the diamond structure. Should the energetic comparison be done at constant density or constant bond length? Figure 24 shows the result of the two calculations[11)], one for each model. Clearly

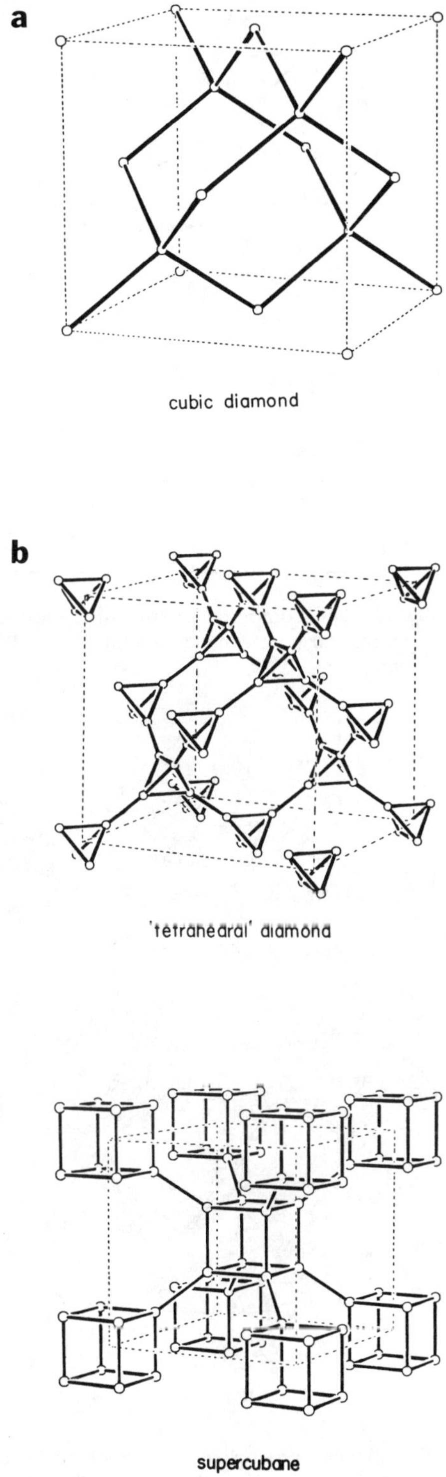

Fig. 21 a–c. The structures of **a)** cubic diamond, **b)** tetrahedral diamond and **c)** supercubane. (**b)** and **c)** are hypothetical.)

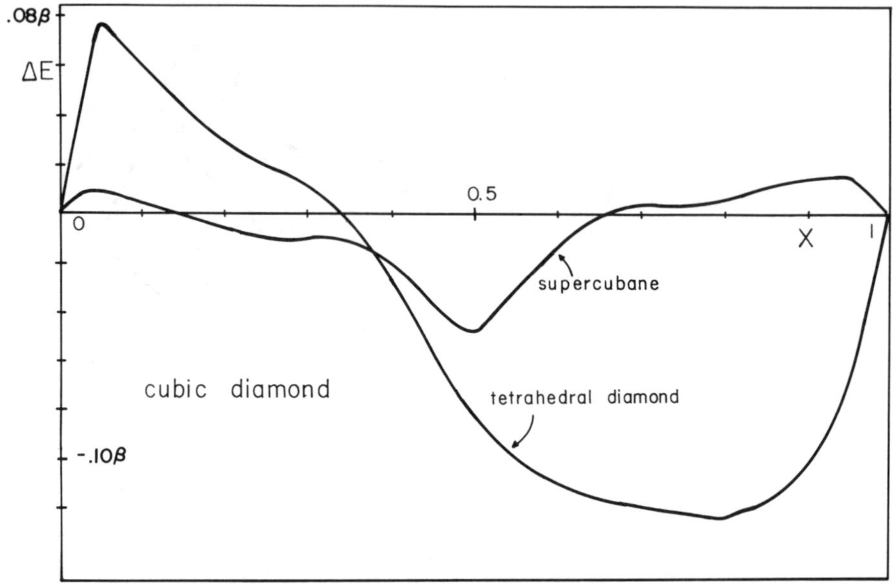

Fig. 22. The computed energy difference curve between cubic diamond and the tetrahedral diamond and supercubane structures

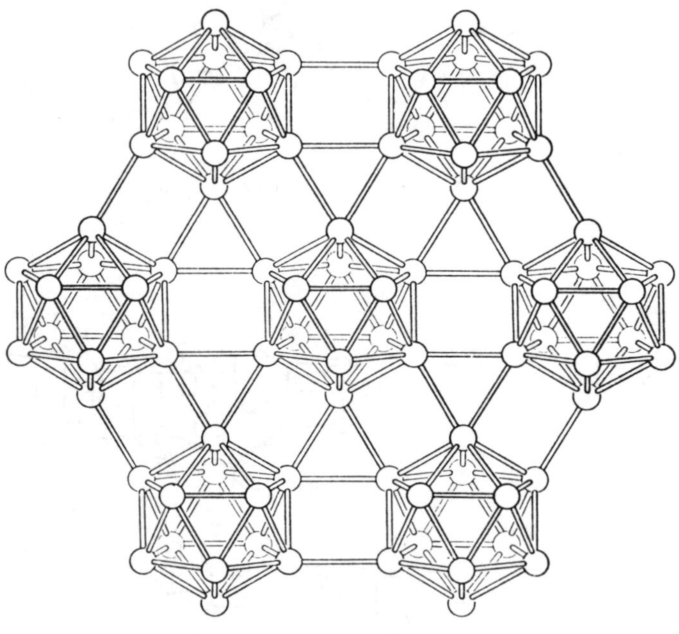

Fig. 23. The structure of rhombohedral boron

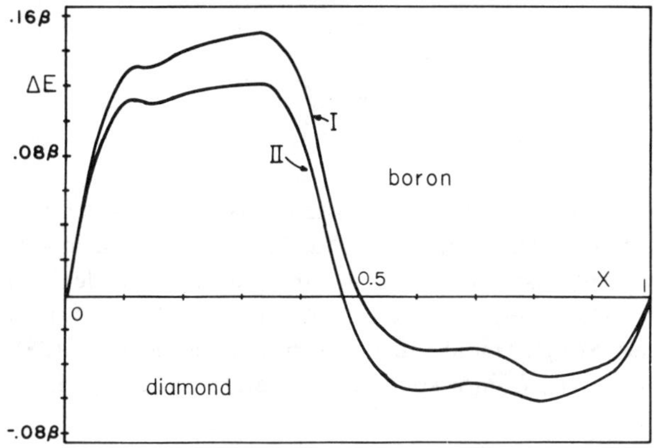

Fig. 24. The computed energy difference curve between cubic diamond and rhombohedral boron

the form of the curve is that expected for a third moment problem but the node has shifted from one model to the other. This arises because of a difference in the second moments of the two structures. The second moment is just the sum of the squares of the H_{ij} interaction integrals between the orbitals of an atom and those of its neighbors. The cubic diamond and rhombohedral boron structures will then have different second moments both because of different bond lengths and different coordination numbers. **25** shows how the energy differences curve changes as a result. (Notice the form of the

25

second moment curve.) One-electron models of this type are generally notoriously inept at making any comments concerning preferred coordination numbers in molecules and solids and so we need to be careful in any problem which involves coordination number differences. However, the general result is clear; the boron structure is favored at the boron electron count because of the presence of three-membered rings and disfavored at the carbon electron count for the same reason.

I. Jahn-Teller Distortions in Octahedral Complexes

The Jahn-Teller theorem[22), 24)] and its structural consequences have long been used to view problems in all branches of chemistry. Since its initial presentation in 1936 many different types of geometrical relaxation have been labeled "Jahn-Teller distortions". It is worthwhile to briefly summarize the present status of this phrase[24)]. As originally stated, the theorem claims that molecules in orbitally degenerate electronic states will distort so as to remove this degeneracy. Usually the asymmetric occupation by electrons of degenerate orbitals will be sufficient to guarantee an orbitally degenerate electronic state, but not always. In the classic case of singlet cyclobutadiene, for example, none of the electronic states arising from the $(e_g)^2$ configuration are degenerate (**26**). However

26

the phase "Jahn-Teller unstable" is used to cover all examples of asymmetric occupation of degenerate orbitals. A further use of the term applies to substituted octahedral complexes, usually with a d^9 configuration. While in the symmetrical MY_6 molecule an orbitally degenerate electronic state (^2E_g) results from the $(e_g)^3$ configuration (**27**), in the cis $MY_4Y'_2$ molecule no such state is possible (**28**) since the point symmetry of the molecule cannot support degenerate representations.

27

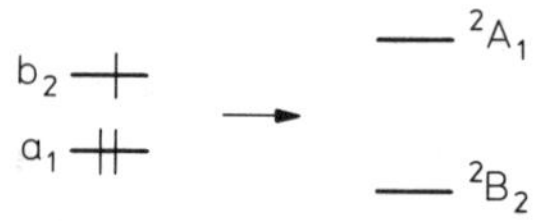

28

Both of these problems have been tackled[24)] by extending the original ideas of Jahn and Teller, via perturbation expansion of the energy of the electronic ground state (0) as a function of a distortion coordinate g. $\mathbf{H}_q$ and $\mathbf{H}_{qq}$ are the first and second derivatives of the electronic Hamiltonian with respect to q

$$E(0) = \langle 0|\mathbf{H}_q|0\rangle + \frac{1}{2}\left[\langle 0|\mathbf{H}_{qq}|0\rangle - 2\sum\frac{|\langle 0|\mathbf{H}_q|j\rangle|^2}{\Delta E_{0j}}\right] q^2 \quad (11)$$

The term linearly dependent upon q is the first order Jahn-Teller one and may be nonzero for degenerate electronic states. It will apply to the d^9 octahedral MY_6 case of **27.** The term in brackets is in two parts. The first is the classical force constant for motion of the molecule along the coordinate q within the charge distribution described by the electronic state $|0\rangle$. The second is the relaxation force constant which allows the electron density to relax during the motion. If there is a low-lying electronic state (j) of the correct symmetry such that the numerator of this expression is significant, then this relaxation contribution may be large. If it is larger than the classical force constant then the term in brackets is negative and the system distorts. In both **26** and **28** there is a low-lying state which can be utilized in this way. In cyclobutadiene where this electronic state is derived from the same configuration as that of the ground state, we call the result a pseudo Jahn-Teller distortion. In the low symmetry d^9 complex of **28** where the electronic configurations of ground and excited states are different, the term second-order Jahn-Teller effect is used[18)]. In the chemical literature the phrase "Jahn-Teller distortion" is usually employed for all three.

However there is a problem with this dicussion for many pseudo-octahedral d^9 complexes. **29** shows the electronic configurations and states expected for a d^9 trans $MY_4Y'_2$ complex (D_{4h}). Whether the a_{1g} or b_{1g} level lies lowest in energy (and hence the $^2B_{1g}$ or $^2A_{1g}$ state) will depend upon the relative σ strength of the ligands, but does not influence our discussion. From the symmetry demands of the relaxation term of Eq. 11 the sym-

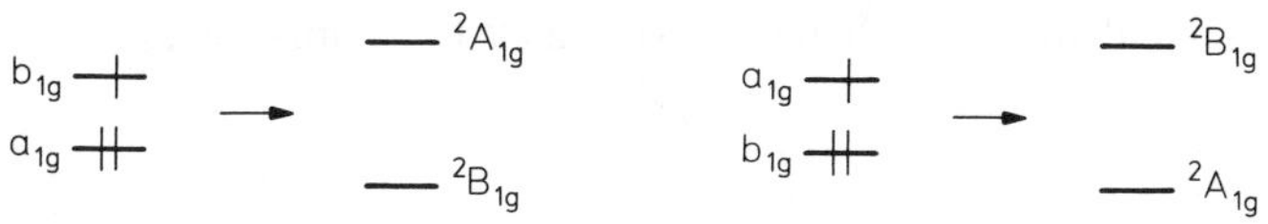

29

metry species of the distortion coordinate q must be $a_{1g} \times b_{1g} = b_{1g}$, irrespective of this level ordering. A distortion of this type (**30**) is indeed found[24)] for $CuCl_2 \cdot H_2O$ and α-$Cu(NH_3)Br_2$. However in a wide range of $Cu^{II}O_2N_4$ complexes with oxygen and nitrogen containing donor ligands an a_{1g} distortion (**31**) is found[24)] which is not compatible

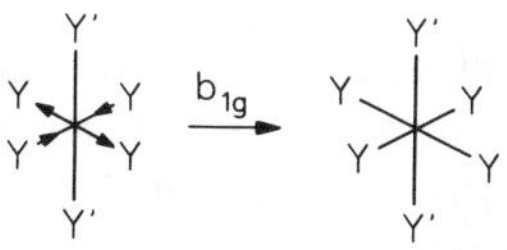

30

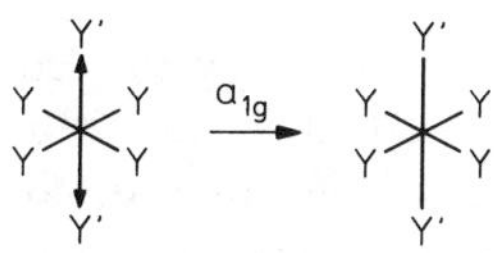

31

with the second-order Jahn-Teller prescription. Let us use the method of moments to tackle this problem.

Let β be the interaction integral between the central atom z^2 orbital and a ligand lying along the z axis. We define two such β values, β_{ax} and β_{eq} for the ligands we will locate in axial and equatorial positions of an octahedron. Remembering (**32**) that the overlap

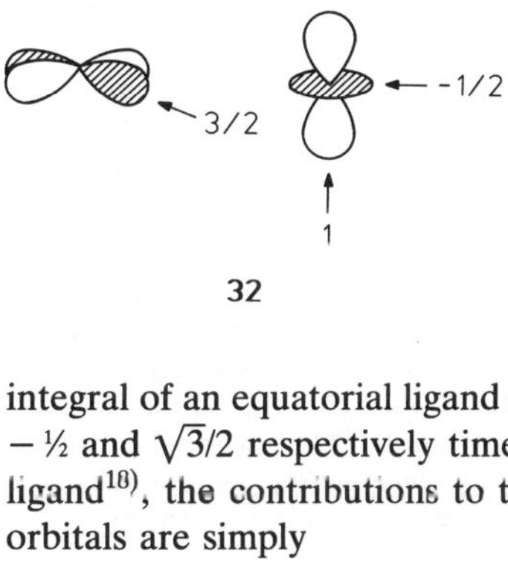

32

integral of an equatorial ligand σ orbital with the collar of z^2 and with a lobe of $x^2 - y^2$ is $-\frac{1}{2}$ and $\sqrt{3}/2$ respectively times the corresponding overlap integral of z^2 with an axial ligand[18)], the contributions to the second moment from walks originating on the metal orbitals are simply

$$\begin{aligned} &\text{for } x^2 - y^2 \quad && 3\beta_{eq}^2 \\ &\text{for } z^2 && 2\beta_{ax}^2 + \beta_{eq}^2 \end{aligned} \tag{12}$$

The total contribution to μ_2 is then

$$4\beta_{eq}^2 + 2\beta_{ax}^2 \tag{13}$$

During the distortion we shall require that the second moment remains constant. (We have described above some of the problems associated with one-electron models and

coordination number, or second moment changes.) This may be simply achieved by writing

$$\beta_{eq}^2 = \beta^2 + x \quad \text{and} \quad \beta_{ax}^2 = \beta^2 - 2x \tag{14}$$

(For small distortions this requirement suggests that the axial ligands will move further than the equatorial ones on distortion a feature borne out by the data[25)] on Jahn-Teller distortions in copper(II) complexes.) The contribution to the fourth moment from the metal based orbitals is simply evaluated as

$$9\beta_{eq}^4 + 4\beta_{ax}^4 + \beta_{eq}^4 + 4\beta_{ax}^2\beta_{eq}^2 \tag{15}$$

which, using the restrictions imposed by the Eq. 14, simply gives a fourth moment contribution of $18(\beta^4 + x^2)$. The Jahn-Teller distortion in these ocathedral complexes is thus a fourth moment problem, the undistorted molecule having the smaller fourth moment. By analogy with our discussion above concerning orbital sharing a simple energy difference plot for the Jahn-Teller process may be constructed as in Fig. 25b. The orbital filling parameter x in this case is just the fractional orbital occupancy of the set of σ orbitals shown in Fig. 25a. Molecules with electron configurations appropriate to the left-hand side of Fig. 25b, with partial ligand orbital occupancy do not exist. Clearly a

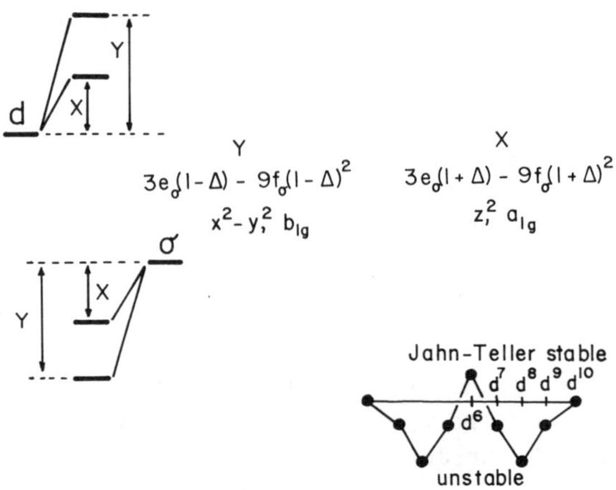

Fig. 25. a) The energy levels of a Jahn-Teller disorted octahedral complex using the angular overlap model ($\Delta = 0$ for the undistorted system) and **b)** the energy difference curve that results. The abscissa here represents the electron count applicable to the σ-orbital collection of **a)**

distortion away from the regular octahedral structure is predicted for molecules with d^7, l.s. d^8 or d^9 configurations. The l.s. (low-spin) d^8 configuration should have the largest distortion since here the fourth moment effect is largest. This is indeed found in practice; l.s. d^8 complexes are never six coordinate and octahedral, but always four coordinate and square planar[26)]. Two axial ligands have been lost altogether. For d^7 and d^9 systems the size of the distortion is variable, but usually the system is described as basically six-coordinate with two long bonds.

Let us now examine the octahedral *trans* $MY_4Y'_2$ molecule and ask using the angular overlap model which distortion out of **30** and **31** will be most favored without any preconceptions induced via the Jahn-Teller theorems. **33** shows the energy levels for

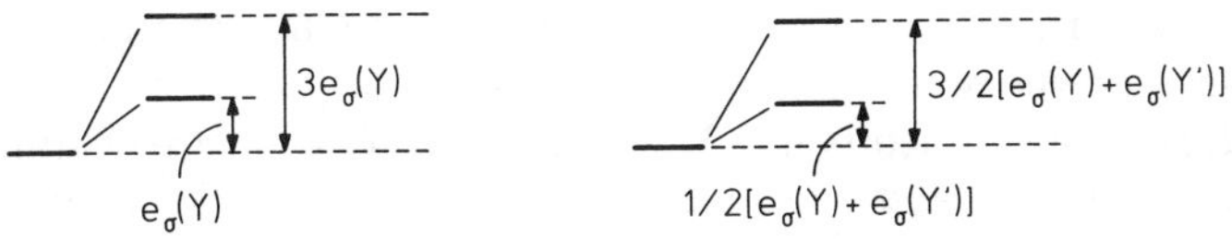

33

square planar MY_4 and MY_2Y_2'. For a d^9 system the MY_4 molecule has a MOSE of $3e_\sigma(Y)$ and the MY_2Y_2' molecule a MOSE of $3/2(e_\sigma(X) + (e_\sigma(Y'))$. In other words elongation of the ligands with the smallest values of e_σ should be the preferred distortion coordinate. The D_{4h} MY_4Y_2' structure will be observed if $e_\sigma(Y) > e_\sigma(Y')$ and the C_{2v} MY_4Y_2' arrangement if the converse is true. Elsewhere we show that this is true in practice[24)]. We can see how this result comes about from the moments approach too. For a *trans* MY_4Y_2' molecule we can take care of the asymmetry of the ligands by setting $\beta_{eq}^2 = \beta^2 + x$ as before but let $\beta_{ax}^2 = \beta'^2 - 2x$. Now the contribution to the fourth moment becomes

$$10\beta^4 + 4\beta'^4 + 4\beta'^2\beta^2 + 12x[\beta^2 - \beta'^2] + 18x^2 \tag{16}$$

This is maximized (i.e., the axial distortion to give eventually a square planar MY_4 unit maximally encouraged) if $12[\beta^2 - \beta'^2] > 0$. This implies that $\beta^2 > \beta'^2$, the same result as before. Perhaps the most important aspect of both the angular overlap or moments approach is that the symmetry of the molecule does not play a role in the treatment. It is just as applicable to *cis* and *trans* MY_4Y_2' species and indeed any low symmetry system as it is to octahedral MY_6 ones. In view of this discussion it is of interest to recall van Vleck's comment in 1939 (quoted in Ref. 22) "It is a great merit of the Jahn-Teller effect that it disappears when not needed".

The distortions we have described above, often attributed to "Jahn-Teller effects" are, of course another example of the busy/idle concept described earlier. The situation is a little more complex here since the two d orbitals involved do not have equal interactions with all ligands but the busy $x^2 - y^2$ and idle z^2 orbital in the axially elongated molecule may readily be compared to the metriotic situation in the octahedral molecule.

J. The Jahn-Teller Effect in Cyclobutadiene and Other Polyenes

34 shows the parameters describing the $p\pi$ interaction integrals for the Jahn-Teller distortion of singlet cyclobutadiene. Again we shall require that the second moment $4(\beta_1^2 + \beta_2^2)$

34

remains constant during this process. The odd moments are of course all zero but the fourth moment may be simply evaluated as

$$\mu_4 = 4(\beta_1^4 + \beta_2^4 + 6\beta_1^2\beta_2^2) \tag{17}$$

which is clearly maximized when $\beta_1 = \beta_2$. So we may view the distortion away from the square toward the rectangular geometry as a way of reducing the fourth moment, energetically unfavorable at the half-filled point. **35** shows a fourth moment plot with relevant electron counts marked on it. With two fewer or two more electrons the instabil-

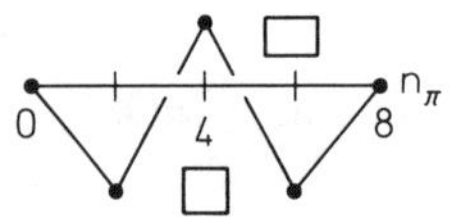

35

ity vanishes. S_4^{2+} is an example of a molecule with two more electrons, stable in the square geometry.

We can apply exactly analogous reasoning to the corresponding bond alternation process in benzene shown in **36.** The fourth moment here is simply

$$\mu_4 = 6(\beta_1^4 + \beta_2^4 + 4\beta_1^2\beta_2^2) \tag{18}$$

also maximized for $\beta_1 = \beta_2$. This suggests that benzene too should undergo an analogous

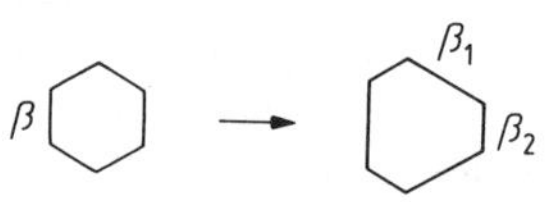

36

distortion, which, of course is not observed. We can see why this might be so. Equations 17 and 18 may be rewritten in terms of the fourth moment contribution per atom as

$$\begin{aligned} &\text{cyclobutadiene} \quad (\beta_1^2 + \beta_2^2)^2 + 4\beta_1^2\beta_2^2 \\ &\text{benzene} \quad (\beta_1^2 + \beta_2^2)^2 + 2\beta_1^2\beta_2^2 \end{aligned} \tag{19}$$

This suggests that the effect should be stronger for cyclobutadiene than for benzene, a result in accord with observation. Resisting such distortions are the energetic preferences of the σ framework. In cyclobutadiene the driving force, via the extra walks of length four around the ring, is large enough to win out, but in benzene it is not. (Computations[27]) on cyclic N_6 or H_6 however show that the distortion should proceed in these molecules).

K. The Peierls Distortion

For an infinite chain of atoms we can imagine several sorts of distortions involving bond length changes. **37** and **38** show two simple ones. **37** is just the bond alternation analogous to the distortions of cyclobutadiene and benzene shown in **34** and **36. 38** is a trimerization. Again we will keep the second moments constant for both distortions and

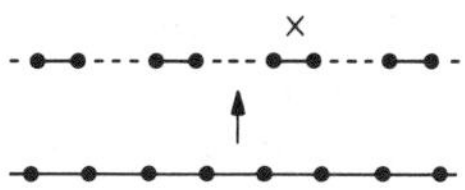

37

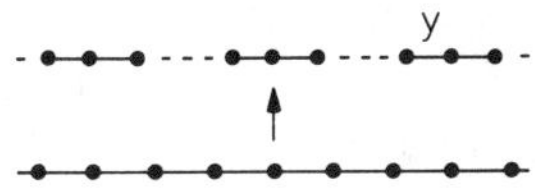

38

Tabel 5. Moments of distorted linear chains

Moment	One atom cell (undistorted)	Two atom cell ($y = 1.1\beta$)[a]	Three atom cell ($x = 1.1\beta$)[a]	Three atom cell ($x = 1.2\beta$)[a]	Three atom cell ($x = 0.8\beta$)[a]
μ_2	$2\beta^2$	$2\beta^2$	$2\beta^2$	$2\beta^2$	$2\beta^2$
μ_4	$6\beta^4$	$5.912\beta^4$	$6\beta^4$	$6\beta^4$	$6\beta^4$
μ_6	$20\beta^6$	$19.471\beta^6$	$16.698\beta^6$	$18.498\beta^6$	$19.410\beta^6$

[a] See **37** and **38** for the location of the linkages with interaction integral equal to $x\beta$ and $y\beta$ respectively

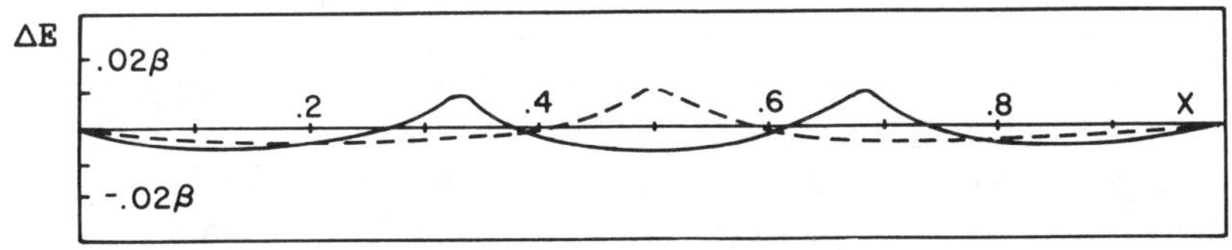

Fig. 26. Computed energy difference curves for Peierls distortions. The solid and dashed lines correspond to dimerization and trimerization respectively **(37, 38)**

ask for which electron counts these two distortions should occur. Table 5 shows the values[11] of the first few moments for these two distorted geometries. The interaction integrals appropriate to the solid linkages of **37** and **38** are set at $x\beta$ and $y\beta$. Notice that just as in the molecular case simple bond alternation is a fourth moment problem and the distortion away from the symmetric structure should therefore occur at the half-filled point. Indeed polyacetylene, $(CH)_n$, is distorted in this way, and a linear chain of hydrogen atoms lined up equidistant from their neighbors clearly distorts to produce a line of H_2 molecules. Figure 26 shows a calculated energy difference curve[11] for the undistorted and alternated structures. For elemental sulfur, with two electrons more than $(CH)_n$ an undistorted chain is predicted, and is observed in the fibrous allotrope. Many examples are known of distortions of this type[2]. In some cases the driving force for distortion is not large and the elastic restoring forces hold the system in a symmetrical arrangement (cf. the earlier discussion on benzene). In some cases observation of a distorted structure is temperature dependent.

For the trimerization, **38,** it is the sixth moment (Table 5) which distinguishes it from the undistorted chain. Since the sixth moment decreases on distortion, the undistorted structure will be more stable at the half-filled point. Figure 26 shows a computed energy difference curve for this distortion too. Notice that the trimerization is predicted to be energetically favorable at the third or two-thirds full band. Here the distortion is the one which relieves an energetically unfavorable sixth moment at these electron counts. Several examples are known. Imagine, for example, a one-dimensional chain of iodine atoms with sufficient electron doping to produce a species $I^{-0.333}$. We may write its atomic configuration as $s^2p_x^2p_y^2p_z^{1.333}$ which implies, if the chain lies along the z axis, filled π bands (x,y) and a two-thirds filled $p_\sigma(z)$ band. A system with such a two third filled band should distort via a trimerization. The result is the well-known I_3^- ion. Another example[28] contains a heteroatom. The $CdCl_2$ structure may be described as a cubic close-packed array of anions with the Cd^{2+} ions occupying one-half of the octahedral holes in a way such that Cl–Cd–Cl slabs are formed. Each chloride is three coordinate, in a pyramidal fashion, by metal. At each metal are three orthogonal Cl–Cd–Cl units. Both σ and π bands of such an extended species are two-thirds full and it is therefore not surprising that the structure is well described by three orthogonal localized systems of this type.

L. The Coloring Problem

A structural problem of considerable interest is associated with the possible distribution patterns of atoms of different types (colors) in a given lattice. Figure 27 shows three examples. What are the electronic factors influencing the stability of the CsCl and CuTi arrangements, both found for transition metal alloys? How should we best stabilize a square singlet cyclobutadiene by attaching donor and acceptor groups to the carbon atoms? Is the *cis* or *trans* arrangement to be favored? Is the most stable graphite derivative (e.g., BN) the structure with all linkages between atoms of opposite type, or are homonuclear contacts possible too? All of these questions may be very briefly summarized quite simply. What are the electronic requirements for stabilizing an ...ABAB... arrangement over an ...AABB... one? A and B may be individual atoms as in Fig. 27a, lines of atoms as in Fig. 27b or planes of atoms as in Fig. 27c.

The simplest way to approach this problem[12] is to enumerate the moments associated with these two possibilities and see how they differ. Initially, let us suppose that the only difference between A and B is a different Hückel α value. With this restriction the moments of a chain of atoms are shown in Table 6a, c. It is quite clear that the two patterns differ at the fourth moment, and that it is the ...AABB... pattern which has the larger moment. Such coloring problems should therefore be controlled by a fourth moment curve, with the ...ABAB... pattern stable at the half-filled point and the ...AABB... pattern stable at the quarter and three-quarter filled points. **39** shows the

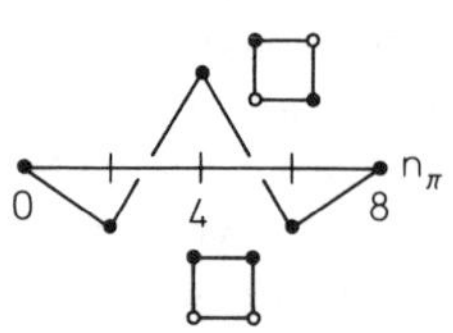

39

results of a simple Hückel calculation for cyclobutadiene, using a model which included just a change in the α values of the carbon atoms to which the groups were attached. Notice that it is indeed the *trans,* or ABAB arrangement which is predicted to be most

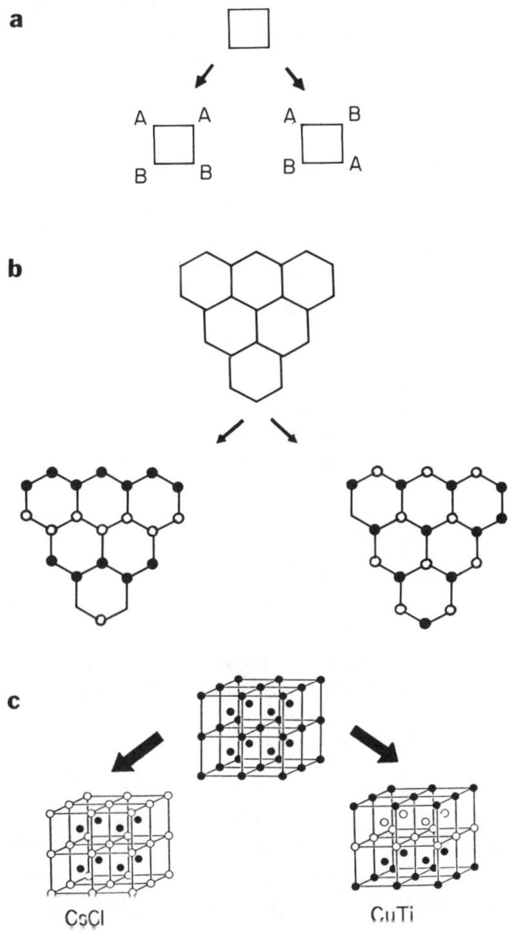

Fig. 27 a–c. Some coloring problems. **a)** Attachment of donors and acceptors to square cyclobutadiene. **b)** Two arrangements of the B and N atoms over the graphite net in BN. (Only the fully alternating one is found in practice). **c)** The CsCl and CuTi structures which result from the body centered cubic lattice

stable. This is exactly in accord with observation. All the known push-pull cyclobutadienes are of this type.

The lowest energy coloring of the no-metal net in CaB_2C_2 (**4**) is open to similar treatment. The net is isoelectronic with graphite. Figure 28 shows[28] the electronic density of states computed for two possibilities. Even by eye it is clear that the structure with no B-B or C-C contacts is the one stable at the electronic configuration appropriate for $B_2C_2^{2-}$ (x for the π manifold is equal to 0.5). Notice that the alternative structure would be metallic at this electron count. (To date there is unfortunately no definitive experimental study on the ordering pattern in this material). With two extra electrons, which give a $B_2C_2^{4-}$ unit ($x = 0.75$) the alternative pattern is computed[29] to be more stable.

The fourth moment curve predicts that the most stable arrangement for BN in the graphite structure should be the one of Fig. 27b which contains no homonuclear contacts. This is indeed the observed arrangement. Computations for the other possibility[29] show that it will not be an insulator at all, but a metal. BN is in contrast to graphite, found as a white insulating material with clearly a good band gap.

Table 6. Moments of AABB and ABAB pattern

	...ABAB...
(a)	all β equal
	$\mu_1 = 2(\alpha_A + \alpha_B)$
	$\mu_2 = 2(\alpha_A^2 + \alpha_B^2 + 4\beta^2)$
	$\mu_3 = 2(\alpha_A^3 + \alpha_B^3 + 6\beta^2(\alpha_A + \alpha_B))$
	$\mu_4 = 2(\alpha_A^4 + \alpha_B^4 + 12\beta^2 + 8\beta^2(\alpha_A^2 + \alpha_B^2 + 8\alpha_A\alpha_B\beta^2)$
(b)	unequal β
	$\mu_1 = 2(\alpha_A + \alpha_B)$
	$\mu_2 = 2(\alpha_A^2 + \alpha_B^a + 4\beta_{AB}^2)$
	$\mu_3 = 2(\alpha_A^3 + \alpha_B^3 + 6\beta_{AB}^2(\alpha_A + \alpha_B))$
	...AABB...
(c)	all β equal
	$\mu_1 = 2(\alpha_A + \alpha_B)$
	$\mu_2 = 2(\alpha_A^2 + \alpha_B^2 + 4\beta^2)$
	$\mu_3 = 2(\alpha_A^3 + \alpha_B^3 + 6\beta^2(\alpha_A + \alpha_B))$
	$\mu_4 = 2(\alpha_A^4 + \alpha_B^4 + 12\beta^2 + 10\beta^2(\alpha_A^2 + \alpha_B^2) + 4\alpha_A\alpha_B\beta^2)$
(d)	unequal β
	$\mu_1 = 2(\alpha_A + \alpha_B)$
	$\mu_2 = 2(\alpha_A^2 + \alpha_B^2 + 2\beta_{AB}^2 + \beta_{AA}^2 + \beta_{BB}^2)$
	$\mu_3 = 2(\alpha_A^3 + \alpha_B^3 + 3\beta_{AB}^2(\alpha_A + \alpha_B) + 3(\alpha_A\beta_{AA}^2 + \alpha_B\beta_{BB}^2))$
(e)	unequal β^a
	$\Delta\mu_2(\text{ABAB} - \text{AABB}) \propto -(\alpha_A - \alpha_B)^2$
	$\Delta\mu_3(\text{ABAB} - \text{AABB}) \propto -9(\alpha_A - \alpha_B)(\alpha_A^2 - \alpha_B^2)$

[a] Assuming $\beta_{ij} \propto (\alpha_i + \alpha_j)$

The ideas sketched out here may be generalized quite readily[12]. It turns out that in general, for any lattice minimizing the number of A-B interactions in a coloring maximizes μ_4. So the structure which is electronically favored at the half-filled point will be the one with the maximum number of heteronuclear contacts, and therefore the minimum number of homonuclear ones.

The B/N analog of polyacene (**40**) has like the parent species, not been made but our arguments would predict that it too will have the B and N atoms in an alternating arrangement as shown. The structure of a heteropolyacene type structure of stoichiometry AlSn is however known[30], **41**, in $Ba_3Al_2Sn_2$. Here the structure contains

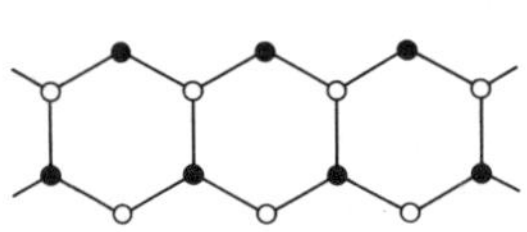

40

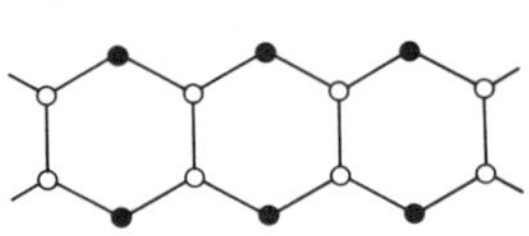

41

homonuclear contacts. Electron counting leads to a value of x = 0.75 for the π manifold of the $(Al^-Sn^-)^-$ sheet, and so the observation of this somewhat unusual arrangement is perfectly understandable. One should ask why a third alternative structure, that containing parallel chains of Sn and Al atoms running along the polymer, is not favored. Here there is the minimum possible number of heteroatom contacts. However the electronega-

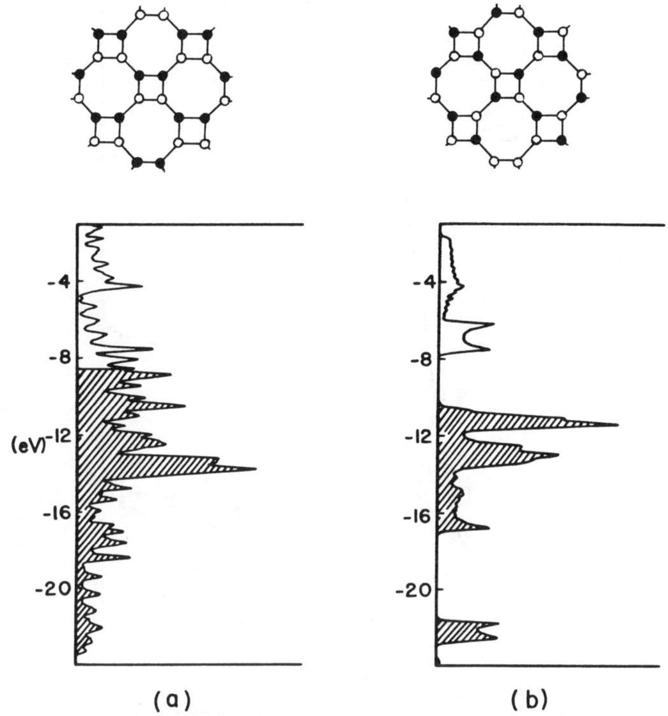

Fig. 28a, b. Computed energy density of states for two possible coloring patterns of the non-metal net (48^2) in CaB_2C_2

tive atoms of systems with such electron counts energetically prefer to reside at sites of low coordination number and this will be an important contributary factor.

Figures 29 and 30 show energy difference curves[12] for the CsCl/CuTi pair of structures of Fig. 27c and the WC/AuCd pair of structures found for alloys of the transition elements with themselves. The CsCl structure is an ... ABAB ... one and the CuTi an ... AABB ... one. The CsCl structure has eight A-B contacts per atom and no A-A or B-B contacts. The CuTi structure has 4 A-B contacts and 4 A-A (or B-B) contacts per atom. The general shape of the energy difference curve is easy to understand in terms of the fourth moment analysis. The AuCd structure has eight A-B contacts and four A-A (or B-B) contacts per atom. The WC arrangement has four of each. Again the shape of the energy difference curve is commensurate with the geometric details of the structure. Satisfyingly too, the results of the computations are in accord with the prevalence of known examples with these structures, shown as dashed lines on the diagrams.

There is however an important feature of these plots that requires comment. The "fourth moment curve" in these calculations has acquired some asymmetry. In the AuCd/WC case where this is most pronounced the WC stability field at high electron counts is quite shallow. The reason[12] for this behavior is quite simple in fact. Our model initially only allowed a change in Hückel α-values, whereas of course the interaction integrals for AA, AB and BB contacts are in general different. If these are allowed to vary then there is a difference in the third moments as shown in Table 6b, d, e. (There is a small

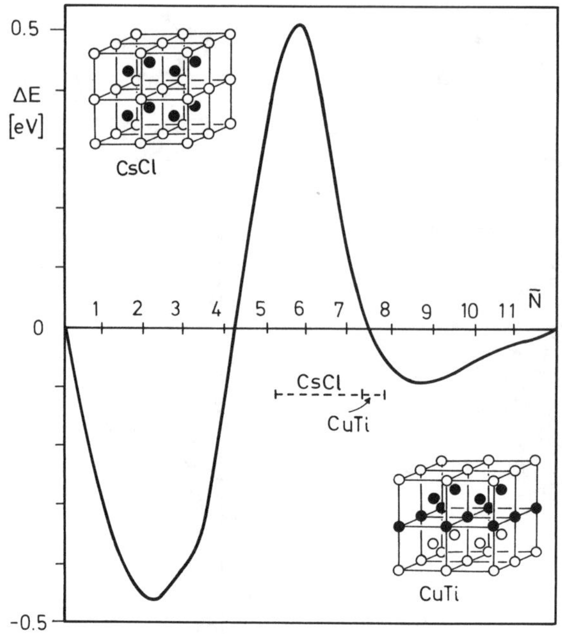

Fig. 29. Computed energy difference for the pair of structure types CsCl and CuTi, for transition-metal transition metal alloys, as a function of the number of d + s electrons. The dashed lines show the electron counts where examples are found experimentally as the lowest energy structures

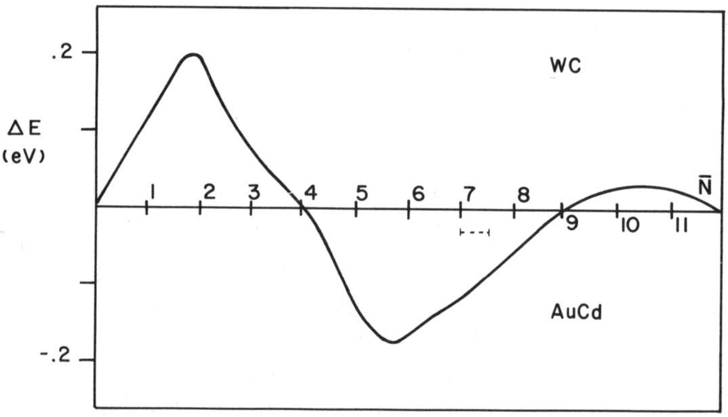

Fig. 30. Computed energy difference curve for the pair of structure types WC and AuCd for transition-metal transition metal alloys. The dashed line shows where examples of the AuCd type are experimentally found

difference in the second moment too). The overall energy difference curve is then a combination of the second, third and fourth moment ones as shown in **42.** If the interaction between A and B is large then the third moment curve will win out and the stability field for the ... AABB ... arrangement at the three-quarter filled point will disappear **(43)**. For interactions between transition metal d orbitals this effect is not too large but for main group systems the third moment may be dominant. Some examples include

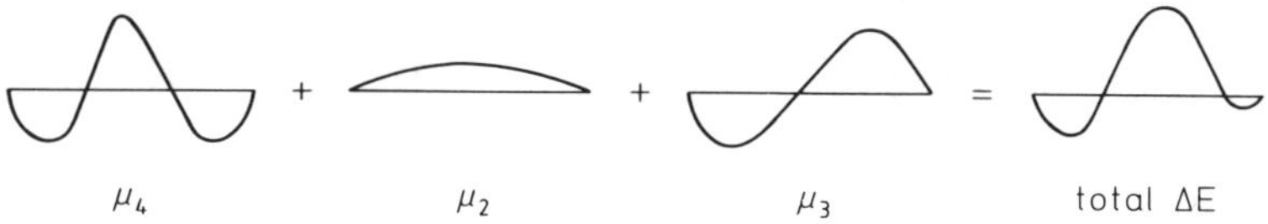

42

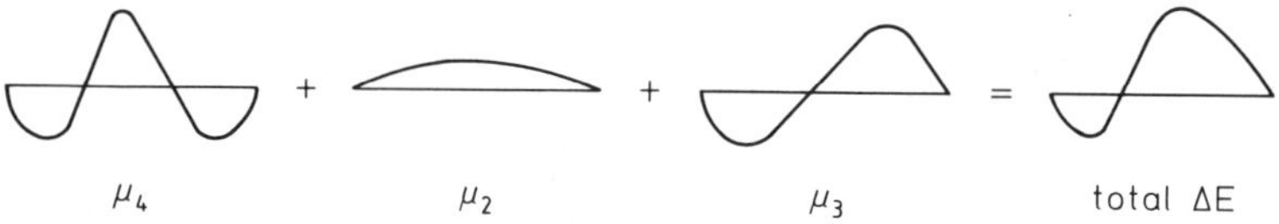

43

square S_2N_2 (found as the ABAB arrangement) and $(SN)_n$ polymer (found as the ...ABAB... arrangement) both of which have x = 0.75.

There is an interesting twist to our rule that the structure with the largest number of A-B contacts is the one favored at the half-full band. Consider two solids where the first coordination shell around each atom is identical, the structural differences appearing at the second nearest neighbor level. **44** shows then how the result is the conversion of walks of length n involving first nearest neighbors into walks of length n + 2 involving second nearest neighbors. So whereas before, at the half-full band, contacts between like atoms were discouraged because they gave rise to an energetically unfavorable fourth moment at this point, like atoms as second nearest neighbors are encouraged at this point because they give rise to an energetically favorable sixth moment. **45** shows the Cu_3Au and Al_3Ti

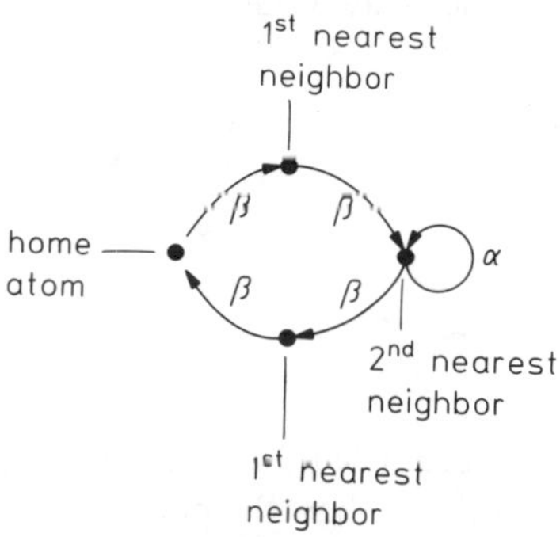

44

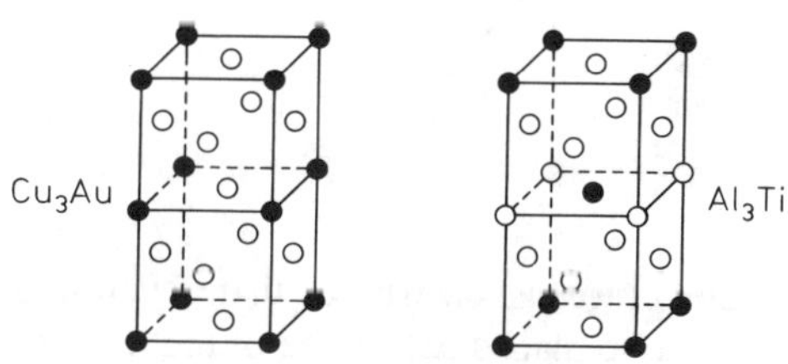

45

structures which are colorings of the fcc lattice. The sixth moment is larger for the Cu_3Au structure since it has the larger number of like second-nearest neighbors and is the structure favored at the half-filled band. Figure 31 shows a computed energy difference curve[31)] for the two structures. The result is a sixth moment curve contaminated by higher moment differences.

In all of the examples we have studied here each of the atoms in the lattice have the same coordination number, and, if we restrict ourselves to modeling the coloring with changes in α values only, have identical third moments. This is not true[12)] of systems where the coordinaton numbers are different. **46** shows an example. We can simply show

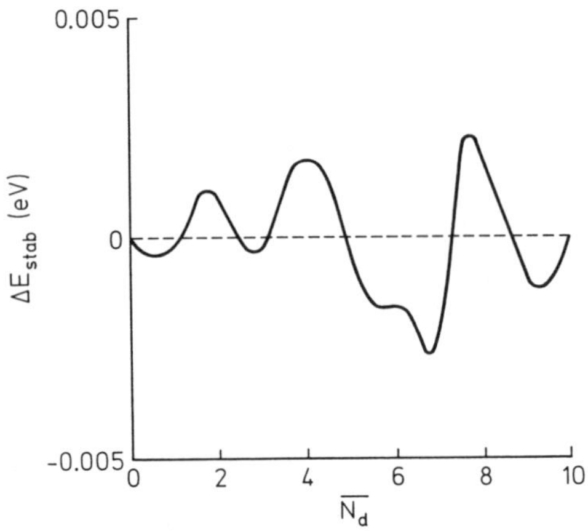

Fig. 31. Computed energy difference curve for the Cu_3Au and Al_3Ti structures. (Redrawn from Ref. 31)

in general, that the part of the third moment which can vary from one coloring to another is given by

$$3 \sum_{i \neq j} \beta_{ij}^2 \alpha_i \qquad (20)$$

If the coordination number (i.e., the sum $\sum_j \beta_{ij}^2$) is the same at each site j then this term remains invariant to the coloring. If the coördination numbers are different then it will become coloring dependent. In order to maximize the absolute value of the third

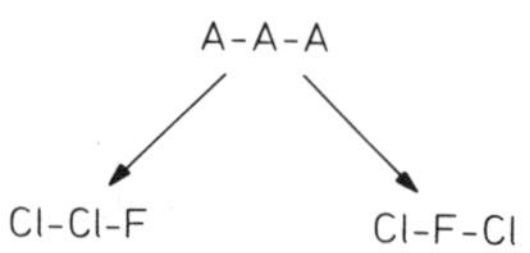

46

moment we can clearly see that the atom of highest electronegativity (largest negative α) should be placed at the site of highest coordination number. We show the moments for the AAB and ABA cases in Table 7. There is also a difference in the second moment if different β values are allowed. If B is more electronegative than A then the ABA arrangement has the higher second moment. This second moment difference has the effect of shifting the node in the third moment curve to higher x, just as in the coordination number problem of **25.** So coloring problems of this type will be determined by an energy difference curve of the type shown in **47.** For the trihalide ions XY_2^- where the valence orbitals are nearly full we need to minimize the third moment and therefore place the most electronegative atom at the end of the molecule as universally found in practice. Br–Br–Cl^- is the isomer found, and not Br–Cl–Br^- for Cl_2Br^- for example. For N_2O with sixteen electrons the stable isomer is NNO and not NON, but for the twelve electron molecule Ga_2O, the observed isomer is GaOGa. The alkali metal oxides (e.g., Li_2O)

Table 7. Moments of molecular AAB and ABA

AAB

(a) All β equal

$\mu_1 = 2\alpha_A + \alpha_B$

$\mu_2 = 2\alpha_A^2 + \alpha_B^2 + 4\beta^2$

$\mu_3 = 2\alpha_A^3 + \alpha_B^3 + 6\alpha_A\beta^2 + 2\alpha_B\beta^2$

(b) Unequal β

$\mu_1 = 2\alpha_A + \alpha_B$

$\mu_2 = 2\alpha_A^2 + \alpha_B^2 + 2\beta_{AA}^2 + 2\beta_{AB}^2$

$\mu_3 = 2\alpha_A^3 + \alpha_B^3 + 4\alpha_A\beta_{AA}^2 + 2(\alpha_A + \alpha_B)\beta_A^2$

ABA

(c) All β equal

$\mu_1 = 2\alpha_A + \alpha_B$

$\mu_2 = 2\alpha_A^2 + \alpha_B^2 + 4\beta^2$

$\mu_3 = 2\alpha_A^3 + \alpha_B^3 + 4\alpha_A\beta^2 + 4\alpha_B\beta^2$

(d) Unequal β

$\mu_1 = 2\alpha_A = \alpha_B$

$\mu_2 = 2\alpha_A^2 + \alpha_B^2 + 4\beta_{AB}^2$

$\mu_3 = 2\alpha_A^3 + \alpha_B^3 + 4(\alpha_A + \alpha_B)\beta_{AB}^2$

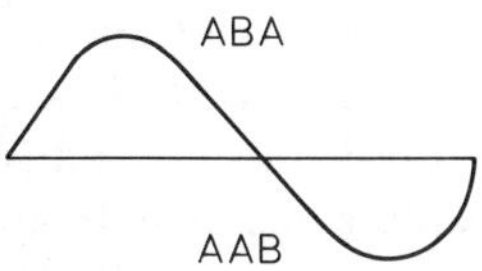

47

with fewer electrons are also found in the symmetrical MOM arrangement. (For a conventional treatment of this problem see Ref. 32). There are many solid state examples too. In the structure of GaS and GaSe which contain three and four coordinate atoms it is the electropositive gallium atoms which are four coordinate and the electronegative chalcogen atoms which are three coordinate. **48** shows an interesting pair of examples commensurate with an electronegativity ordering N > S > As.

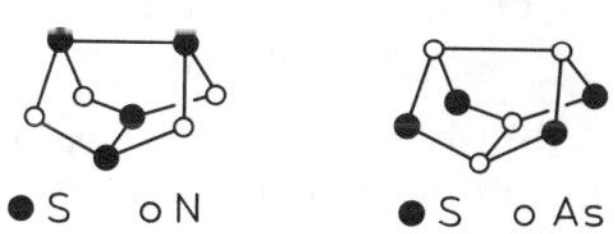

48

Perhaps the majority of systems which are known are found in the ... ABAB ... arrangement. This includes the well known ZnS, NaCl and CsCl structure types and almost certainly occurs as a result of the fact that many stable systems occur when the collection of bonding orbitals are occupied but the corresponding antibonding orbital collection is empty. Such a situation often corresponds to a half-filled band (x = 0.5). ... AABB ... stackings are usually regarded as somewhat unusual by comparison. For example, note the ZrCl structure with its proximal layers of metal atoms.

M. Three Membered Rings in Molecules

The idea of stable three membered rings for systems with less than a half-filled collection of orbitals, as indicated by the third moment curve of Fig. 6 has bearing on the structures of molecules too. Figure 32a shows examples of known, isoelectronic molecules which contain three membered XHH rings. (The first two have been structurally characterized, the last has only been seen in a mass spectrometer). In each case there are a total of two electrons in the set of molecular orbitals generated by the ring. So x = 0.33. We can see this quite clearly in **49.** The frontier orbitals of H^+, $Cr(CO)_5$ and CH_3^+ are empty but two electrons are provided by the H_2 moiety.

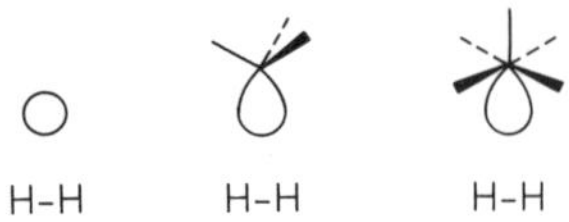

49

The geometries of the first two molecules have been experimentally characterized as shown[33,34]. There is another known species, $[W(CO)_3(PR_3)_2H_2]$, isoelectronic with the second which has been characterized by X-ray and neutron diffraction[35]. The geometry we show for CH_5^+ is the lowest energy one determined by *ab initio* methods[36]. Three membered rings with two pairs of electrons (x = 0.67) are predicted to be unstable on our approach. Consider the structures of the systems in Fig. 32b. Although H_3^- and CH_5^- are calculated[37] to be unbound, the geometry of the latter is the S_N2 transition state and is found in the isoelectronic molecule PF_5. There is no evidence for a three membered ring analogous to that in CH_5^+. For square planar transition metal complexes with the l.s. d^8 configuration a filled z^2 orbital points at an approaching H_2 molecule. In the oxidative addition process that results, an open three-center HMH system is created. Thus closed electron poor three-center bonding but open electron-rich three-center bonding, long a recognized structural feature of inorganic compounds is a part of a larger structural problem.

There are some interesting observations too concerning the structures of polyhedral molecules. Very often they are "electron deficient" in the sense that there are fewer than two electrons for each close contact. The heavy atoms forming the skeleton of the molecule may be either main group atoms (for example in the boranes and carboranes) or transition metal atoms (metal cluster compounds) or both (metallocarboranes). **50** shows the structures expected from Wade's rules for five atom polyhedral molecules with six, seven and eight pairs of skeletal electrons. There are a total of fifteen skeletal orbitals

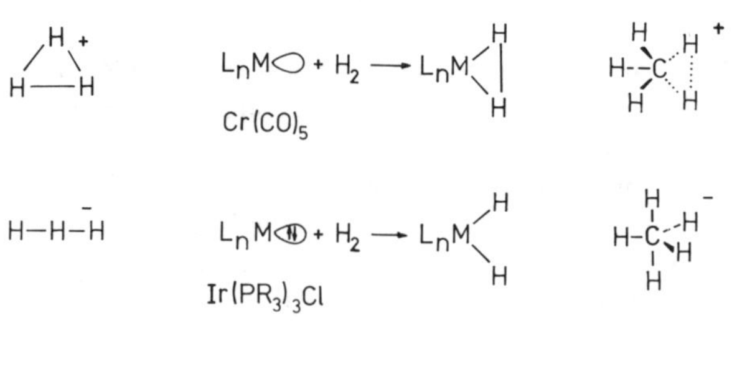

Fig. 32 a, b. Two sets of isoelectronic molecules which show **a)** closed electron poor three center bonding **(49)** and **b)** open electron rich three center bonding

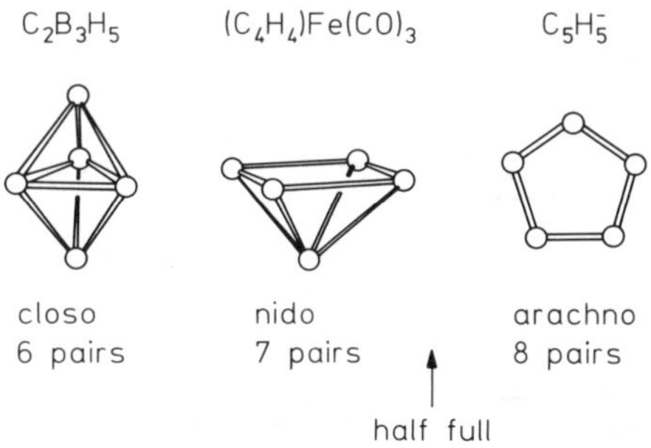

50

for a molecule with five heavy atoms and so the half-full point lies between the last two examples. Notice the large number of three membered rings for the *closo* molecule. In fact all of these electron deficient molecules are based on deltahedra, i.e., polyhedra whose faces are made up of triangles – three membered rings. As the number of electrons increases in the series of molecules of **50** so the number of three-membered rings decreases. Just after the half-filled point notice the single five membered ring in the structure and the absence of three membered rings.

N. The Woodward-Hoffmann Rules

Figure 33 shows molecular orbital correlation diagrams for the disrotatory (**51**) and conrotatory (**52**) electrocyclic ring opening pathways for cyclobutene. With a total of four

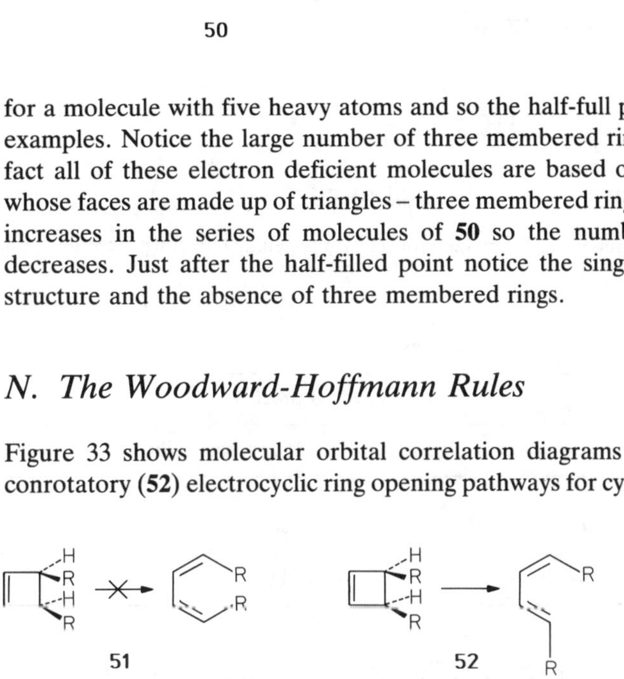

51 **52**

electrons involved with this set of orbitals the disrotatory route is clearly the one which is thermally favored. Diagrams such as these were constructed[38, 39] using the principle of orbital symmetry conservation. In the disrotatory pathway a mirror plane is preserved along the entire reaction coordinate and in the conrotatory pathway a twofold rotation axis is preserved. All of the orbitals of the reactant and the product may be classified either as symmetric or antisymmetric with respect to these symmetry elements. Since orbitals of the same symmetry may not cross along the reaction pathway, and since orbitals at the left-hand side of the diagram must correlate with orbitals of the same symmetry at the right-hand side, the form of the diagram is derived quite simply. Conclusions such as these presented in an easily assimilated form for cycloaddition reactions and electrocyclic ring closures, for example, revolutionized the way organic chemists thought about their subject. A general result which came from these studies was that ring closures/openings proceeded thermally by a disrotatory route if there were 4 N + 2 electrons in the collection of frontier orbitals of the problem and by a conrotatory route if there were 4 N electrons (Table 8). Analogously cycloaddition reactions were thermally forbid-

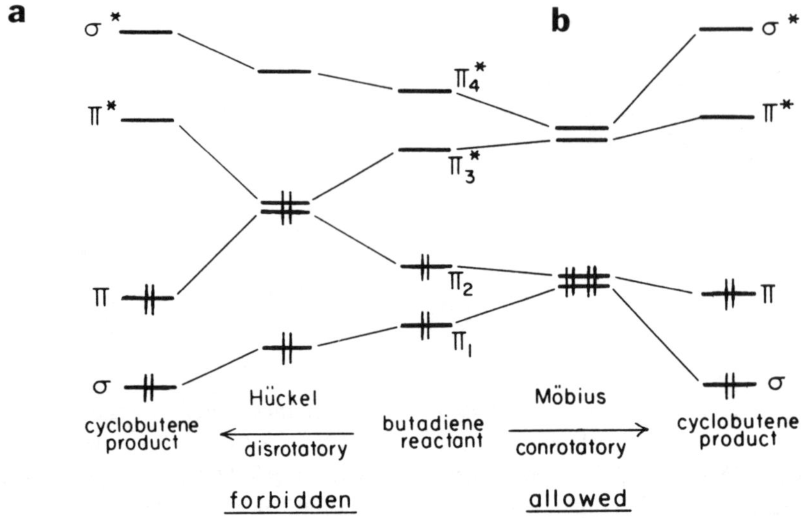

Fig. 33 a, b. Orbital correlation diagrams for an electrocyclic ring opening reaction proceeding by **a)** the disrotatory or **b)** the conrotatory route

Table 8. Selection rules for ring closures and openings

No. of π-electrons in cyclic members	Favored process	
	Thermally	Photochemically
4n (4n − 1)	con	dis
4n + 2 (4n + 1)	dis	con

den if there were 4 N electrons in the problem but allowed if there were 4 N + 2 electrons (Table 9). For carbon chemistry these electron counts could readily be identified with rings of various sizes. Table 8 also contains for completeness the rules for photochemically induced reactions which however we will not discuss in this article. We can use the moments method to provide added insight into this problem and generate a picture very similar to that of Zimmermann[40] who used the Hückel/Möbius concept to study these reactions.

53 and **54** show the geometrical arrangements for the four frontier orbitals of the reactions of **51** and **52.** There is one extremely important difference. For **52,** however we

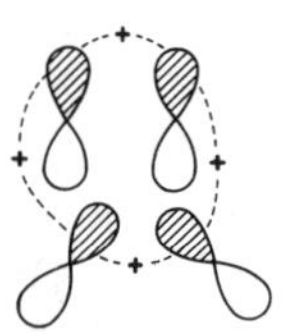

53

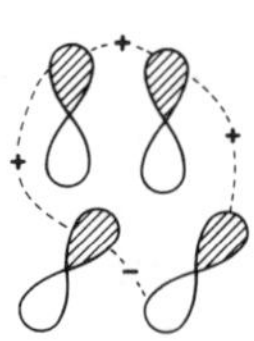

54

Table 9. Selection rules for cycloadditions

Type	Allowed processes Thermal[a]				Photochemical				
	(number of π-electrons involved in each open chain system)								
		M	N			M	N		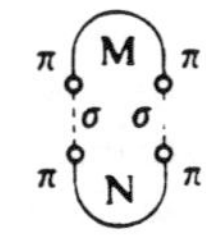
2π → 2σ		4	2			2	2		
		6	4			4	4		
		8	2			6	2		
	L	M	N		L	M	N		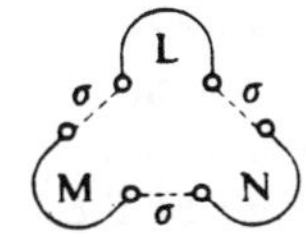
3π → 3σ	2	2	2		4	2	2		
	4	4	2						
	6	2	2						
	K	L	M	N	K	L	M	N	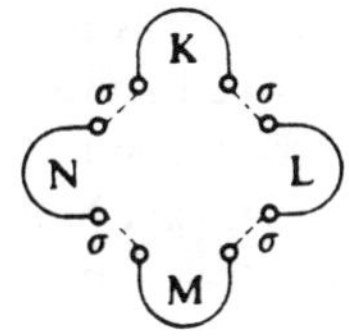
4π → 4σ	4	2	2	2	2	2	2	2	

[a] Note that in all the thermally allowed processes the sum of the π-electron involved in the cycloaddition is (4n + 2)

choose the signs for the lobes of the p orbitals there will always be either an even or zero number of sign inversions between these orbitals. Such an arrangement is called a Hückel system. For **51** there will always be an odd number. This arrangement is called a Möbius system[41)]. How does this affect the sizes of the various moments of the two problems? Figure 34 shows the walks leading to μ_4 for the two sorts of rings, the first moment at which there will be a difference. Notice for the case corresponding to the Möbius system the fourth moment has been reduced compared to the Hückel situation since now the weight of the walk which goes all the way around the ring enters with a negative sign. This is a direct consequence of the odd number of sign inversions on moving around the ring in **51.** So the energy difference curve between Hückel and Möbius four membered rings will be determined by the fourth moment. Since the Möbius four membered ring has the smaller fourth moment it will be the one which is favored at the half-filled point. Thus the conrotatory ring closure reaction is the one with the energetically favored transition state, and therefore is the preferred route.

Consider now a six membered ring as the transition state in the ring closure of **55.** Now the Möbius six membered ring (the conrotatory pathway) has a smaller sixth

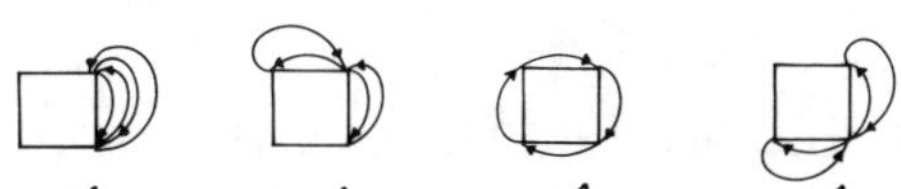

					total
Hückel	$2\beta^4$	$2\beta^4$	$2\beta^4$	$2\beta^4$	$8\beta^4$
Möbius	$2\beta^4$	$2\beta^4$	$-2\beta^4$	$2\beta^4$	$4\beta^4$

Fig. 34. The walks associated with four membered rings of Hückel and Möbius types

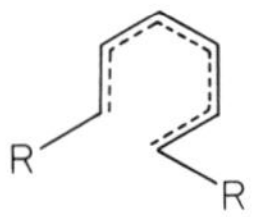

55

moment than does the Hückel six membered ring (the disrotatory pathway). This arises of course since, in an analogous fashion to Fig. 34, the walk around the ring enters with a negative sign. However since large sixth moments are stabilizing at the half-filled point it is now the disrotatory pathway which is energetically favored. Such results are in complete agreement with the results described earlier in Tables 8 and 9. One important comment remains to be made. Notice that, in contrast to the original Woodward-Hoffmann formulation, no symmetry arguments have been used at all in the moments treatment of the problem. In this sense it has analogies with our discussion of the "Jahn-Teller" instabilities of an earlier section, where symmetry arguments (which sometimes worked and sometimes did not) were replaced by an all-encompassing model based on orbital connectivity alone. So in cases where there is no symmetry the moments approach or Hückel-Möbius concept scores over the conventional orbital correlation method. An example, mentioned by Zimmerman[40] is the suprafacial or antarafacial hydrogen or alkyl group migrations. **56** and **57** show the intermediate species involved in 1,7 hydrogen

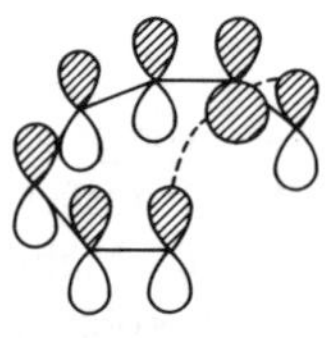

suprafacial migration
(Hückel)

56

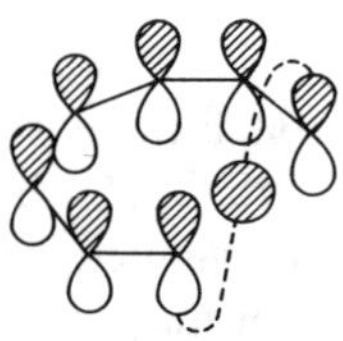

antarafacial migration
(Möbius)

57

migrations. Since the delocalized network contains eight electrons and an eight membered ring, a Möbius system is favored. (The form of the eighth moment curve is easily inferred from Fig. 6). This ensures motion of the hydrogen from the bottom to the top of the π system. Table 10 shows the rules for these systems.

The Hückel-Möbius concept may also be applied to the allene problem we described earlier. Reference 40 shows the "walks" or orbital overlap pathways which distinguish the two isomers.

The energy levels of Hückel and Möbius orbital arrays are easily generated[41]. (In Ref. 14 we show in detail how the orbital energies of Möbius cyclobutadiene are obtained

Tabel 10. Selection rules of [l,j] sigmatropic shifts

$\sum(1 + j)$	Thermally allowed	Photochemically allowed
4n	antara	supra
4n + 2	supra	antara

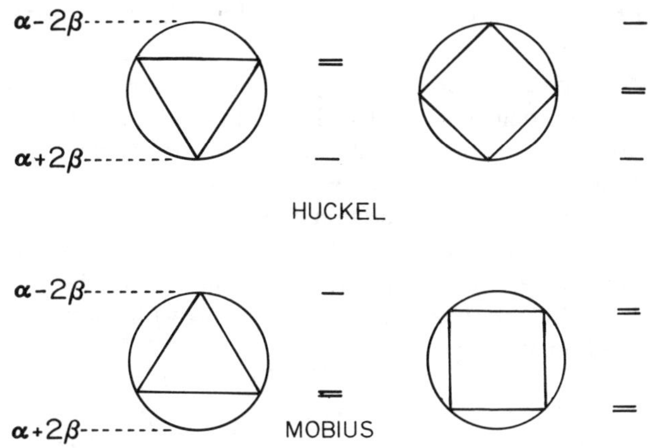

Fig. 35. Frost circle mnemonics for Hückel and Möbius energy levels illustrated with those for three and four membered rings

Table 11. Energetics of Hückel and Möbius systems

No. of electrons	Type of system	
	Hückel	Möbius
4N + 2	Allowed (aromatic)	Forbidden (antiaromatic)
4N	Forbidden (antiaromatic)	Allowed (aromatic)

by use of the moments method). General expressions for the energies are given in Eqs. 21 and 22. k is the level index which runs from 0 to n−1.

Hückel $$e = \alpha - 2\beta\cos(2\pi k/n) \qquad (21)$$

Mobius $$e = \alpha - 2\beta\cos[(2k + 1)\pi/n] \qquad (22)$$

n is the ring size. Mnemonics for the energies are shown in Fig. 35. Notice that levels for cyclobutadiene are the orbital patterns found in the intermediate species of Fig. 33. It is clear to see that the number of electrons which are required to give closed shell, aromatic behavior, in the two are markedly different. Table 11 shows some conclusions which may readily be drawn from such a figure and universally describe all these processes. Notice how cleanly this fits in with the moments treatment. As examples, **58** and **59** show energy

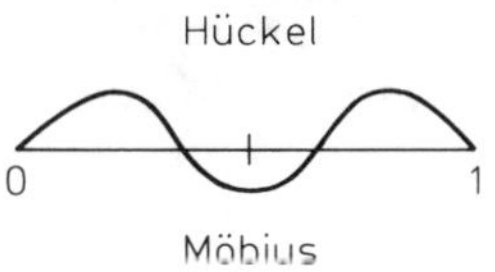

58

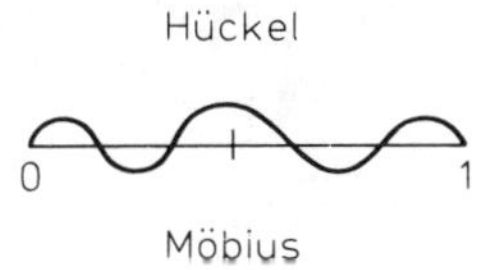

59

difference curves for the Hückel and Möbius four and six membered ring systems we have described earlier.

O. Expansion of the Secular Determinant

The moments method clearly has close ties with the way the secular determinant of the orbital problem is handled. Let us first compare the moments of the energy level spectrum obtained by use of the angular overlap model (a perturbation approach) with those obtained by the use of the walk method. The first two nonzero moments are reproduced well by the angular overlap model (Fig. 36 and Table 12) along with some higher terms. The reader may see quite clearly however that the term in f_σ has to be included in order to accurately recover μ_4. This is an interesting observation in view of our earlier comments linking fourth moment problems with energies dependent upon f_σ only at half-filled point.

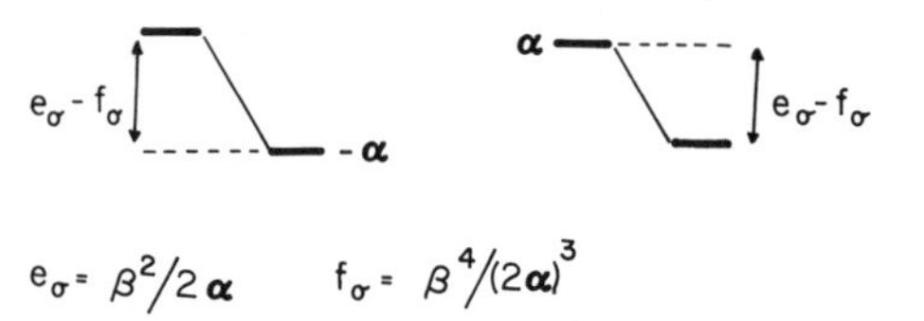

Fig. 36. Angular overlap energy levels for a two orbital problem. In order to ensure the odd moments of the system are zero the Hückel α values of the two basis orbitals are set at ± α

Table 12. Moments and the AOM expansion of the orbital problem

	From Fig. 36		From Walks
	e_σ only	e_σ and f_σ	
μ_2	$2(\alpha^2 + \beta^2)$	$2(\alpha^2 + \beta^2) + \mathrm{Tm}(\beta^6/\alpha^4)$	$2(\alpha^2 + \beta^2)$
μ_4	$2(\alpha^4 + 3/2\beta^4 + 2\alpha^2\beta^2) + \mathrm{Tm}(\beta^6/\alpha^2)$	$2(\alpha^4 + \beta^4 + 2\alpha^2\beta^2) + \mathrm{Tm}(\beta^8/\alpha^4)$	$2(\alpha^4 + \beta^4 + \alpha^2\beta^2)$

We may also study the effect of a linkage induced by a change in the nature of an atom two bonds away. The simplest model would be the A–B–A–B chain shown in **60.**

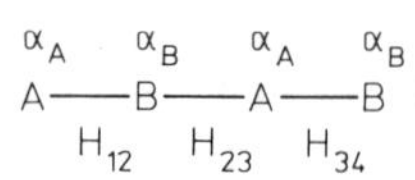

60

We can examine the change in the stabilization energy associated with the linkage between atoms 3 and 4 as a result of a change in the H_{12} interaction parameter. **61** shows the energy level diagram for such a unit. We shall be interested in the sum of the

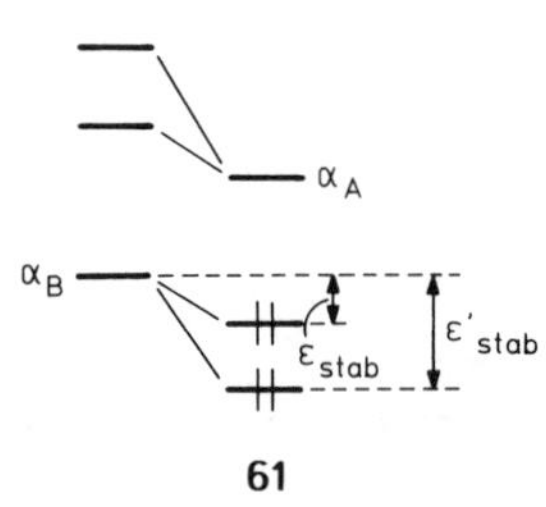

61

stabilization energies of the two lowest orbitals, $\varepsilon = \varepsilon_{stab} + \varepsilon'_{stab}$. The secular determinant for such a problem is just

$$\begin{vmatrix} \alpha_A - E & H_{12} & 0 & 0 \\ H_{12} & \alpha_B - E & H_{23} & 0 \\ 0 & H_{23} & \alpha_A - E & H_{34} \\ 0 & 0 & H_{34} & \alpha_B - E \end{vmatrix} = 0 \tag{23}$$

We may use simple perturbation theory to evaluate ε. Correct to fourth order in H_{ij} this is

$$\varepsilon = \frac{H_{12}^2 + H_{23}^2 + H_{34}^2}{\Delta E} - \frac{H_{12}^4 + H_{23}^4 + H_{34}^4 + 2\,H_{12}^2H_{23}^2 + 2\,H_{23}^2H_{34}^2}{(\Delta E)^3} \tag{24}$$

Of interest is that the fourth order term contains elements which connect adjacent linkages (e.g., $H_{12}^2H_{23}^2$) but contains no terms such as $H_{12}^2H_{34}^2$, which would connect the two terminal linkages of the unit. In fact in order to be able to include interactions joining atoms n linkages away, the determinant needs to be expanded up to order 2n, the number of steps it takes to go from one atom to another and back. For the example of **60** this is six steps and there is indeed a contribution to the energy, ε, in sixth order of

$$\frac{H_{12}^2H_{23}^2H_{34}^2}{(\Delta E)^5} \tag{25}$$

So there is a direct correlation between the order of a perturbation type expansion of the secular determinant problem and the index of the lowest moment which can influence a structural problem. The energetic importance of these higher order terms clearly decreases as 2n increases via the denominator of the perturbation expansion. This relationship between the moments approach and more traditional viewpoints is therefore quite a strong and interesting one.

IV. Some More Difficult Structural Problems

A. *The Crystal Structures of the Transition Metals*

This is not a particularly difficult structural problem but the way the moments contribute to the energy difference curve is difficult to visualize and so it is included here. Figure 37 shows a computed energy difference curve[11] using a d orbital only model for the three common structures adopted by these elements. Such energy difference curves have been in the literature for many years[42]. This one was generated using the Hückel method. There is an interesting observation concerning electron counting for these systems. The way the s orbitals mix into the d band results in there being 1–1.5 s electrons per metal

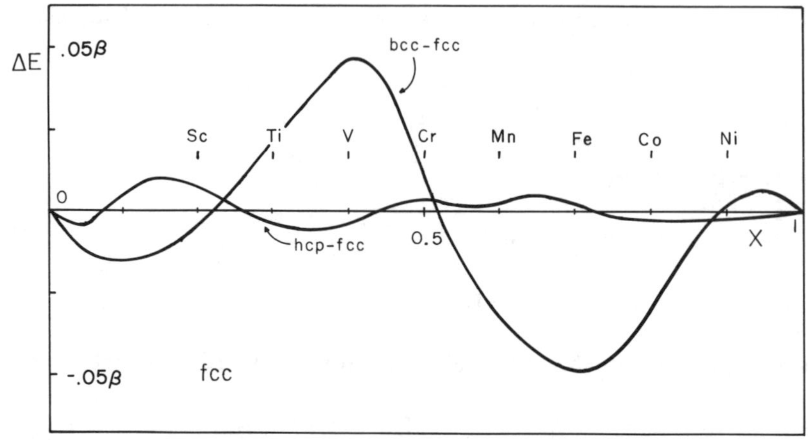

Fig. 37. Computed energy difference curve for the three crystal structures adopted by the majority of the transition elements

atom across the series. The half-filled d band therefore occurs at Group 6. This is reflected, amongst other things by a maximum in the cohesive energy at this point for the transition metals. Bearing this in mind, the calculated energy difference curve reproduces trends quite well. Groups 3 and 4 have the hcp structure, 5 and 6 the bcc structure, 7 and 8 the hcp structure (except for the magnetic first row metals of these groups) and, 9 and 10 the fcc structure (except for magnetic cobalt with the hcp arrangement). The only major problem with the computed curve is the prediction of the bcc structure at high band fillings, which is not observed.

To a transition metal chemist used to understanding the gross features of transition metal complexes using a σ only model, the obvious first step in looking at this problem is to ignore π and δ interactions between the metal atoms. The result, shown in Fig. 38, is

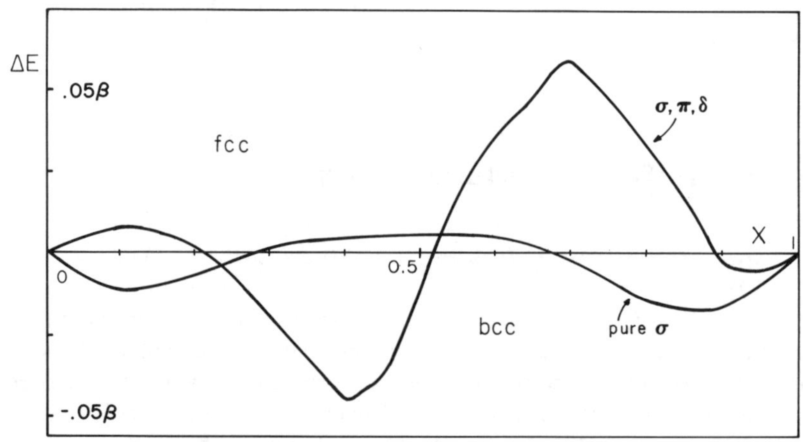

Fig. 38. Computed energy difference curve for the transition element crystal structures using a σ only model, to be compared with the calculation in which all interactions were included

dramatically different from that of Fig. 37. Two things are clear. First, π bonding (and perhaps δ bonding) is of vital importance in structural differentiation. Second, the σ-only energy difference curve between bcc and fcc structures is clearly controlled by a fourth moment difference. The first is difficult to see. We have not been able to present the importance of π type interactions in any pictorial detail. Several walks of different lengths are important. **62** shows a walk of length three which, in the bcc structure,

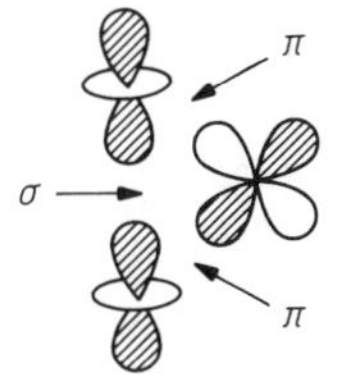

62

contains two π steps and one σ step. In perhaps a way more severe than that encountered in describing the fifth moment difference for the p block heavy elements, the enumeration of the walks (when drawn out in geometrical terms) leaves us with no clear picture of what is going on. Understanding the symmetry associated with the energy difference curve for the σ-only model is somewhat simpler.

The curve has close to mirror symmetry about the x = 0.5 position. This implies that the density of states plots for the two structures have a silimar symmetry. **63** shows that

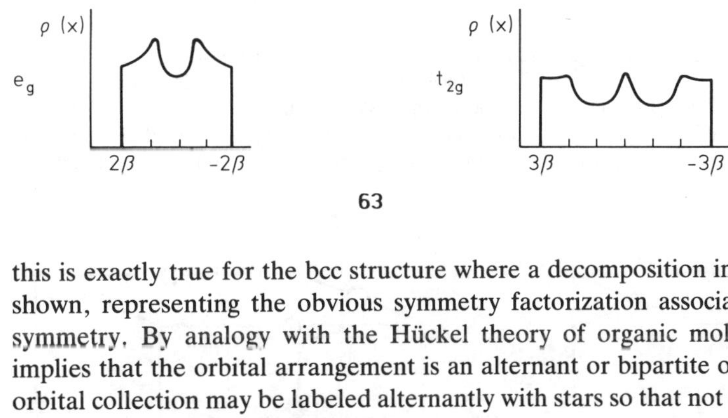

63

this is exactly true for the bcc structure where a decomposition into e_g and t_{2g} bands are shown, representing the obvious symmetry factorization associated with a cubic local symmetry. By analogy with the Hückel theory of organic molecules[43)] such a result implies that the orbital arrangement is an alternant or bipartite one. In other words the orbital collection may be labeled alternantly with stars so that not two starred orbitals, or unstarred orbitals are adjacent. (This is tantamount to saying there are no odd membered walks in the structure). In fact, as indicated in Fig. 39, there is no overlap between "t_{2g}" and "e_g" orbitals at all. In addition the t_{2g} orbitals only interact with orbitals on the first, and the e_g orbitals with orbitals on the second nearest neighbor atoms. So we can therefore envisage two interpenetrating bipartite networks within the crystal. These simply generate a symmetrical density of states plot. For the fcc crystal such a separation is not exactly perfect (Fig. 40). For example the σ overlap between two z^2 orbitals located on first nearest neighbors, clearly zero for the cube-based atoms of **64** in the bcc arrangement, is nonzero (but not very large) for the case of the fcc arrangement, **65.** The result is that the fcc density of states plot is slightly asymmetric. This shows up as the lack of an

64

$f(\sigma, \theta, \varphi) = 0$
bcc
$\theta = 54.73°$

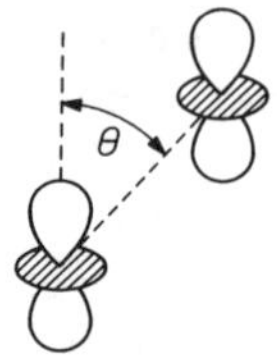

65

$f(\sigma, \theta, \varphi) \neq 0$
fcc
$\theta = 45°$

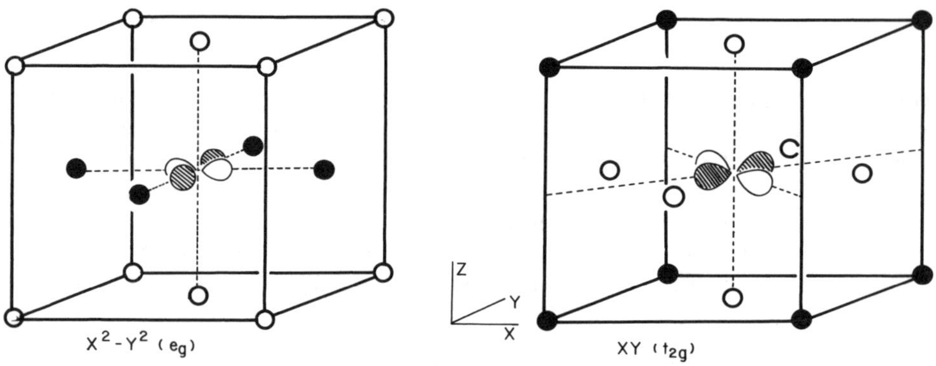

Fig. 39. Overlap between d orbital neighbors in the bcc structure. For clarity second nearest neighbors have been moved in to sit in the centers of the faces of the cube. Empty circles represent atoms which lie on a σ nodal plane; filled circles are those which don't. The $x^2 - y^2$ and xy orbitals are taken as representatives of the e_g and t_{2g} sets respectively. For the case of e_g orbitals there is no overlap with first nearest neighbors, and no overlap of the t_{2g} orbitals with second nearest neighbors

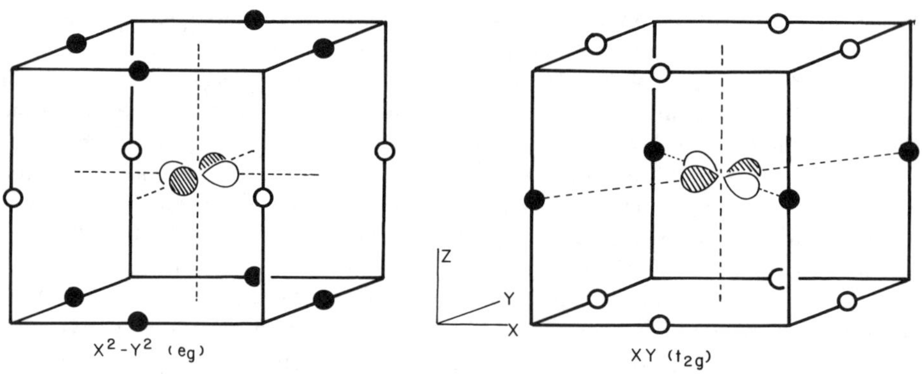

Fig. 40. Orbital overlap in the fcc structure. (See Fig. 40 for conventions). The nearly bipartite character of the σ-only model in fcc can be anticipated in part by noting that four of the five orbitals (including the two here) when considered singly one bipartite

exactly perfect mirror symmetry in the σ-only plot of Fig. 38. However, although we can see how an even moment difference controls the form of this curve, it is difficult to pinpoint that geometrical feature which identifies the order of the difference and why it is the bcc structure with the larger fourth moment. Numerical evaluation of the walks and their weights is straightforward but a one-sentence reason for the calculated energy difference curve of Fig. 37 is at present beyond us.

B. Bond Angle Changes

It is easy to see that in general for atom 1 to be aware of the angle made by atom 3 with the 1–2 bond in **66,** it is a walk of length four which gives the lowest moment different in

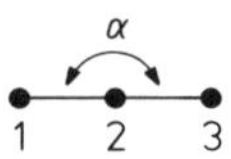

66

linear and bent geometries. We have already seen in Sect. III.D how such a fourth moment curve well reproduces the geometries of the dihydrides of first row elements in accord with Walsh's rules. Figure 41 shows the $\Delta E_4(x)$ curve from Fig. 6 labeled with examples of known first row triatomic molecules. Notice how the crossing point of this curve exactly separates sixteen electron linear, and seventeen electron bent species. There is however, a problem with F_3^-. With twenty-two electrons it is predicted to be nonlinear, but the experimental data (mutual exclusion of IR and Raman bands) for the matrix isolated species indicate a linear geometry[44]. The experimental result is supported by ab initio computations on the ion.

We must not be taken in however by the apparent success shown in Fig. 41. The Walsh numbers, the points at which AB_n molecules distort from linear to bent or from planar to pyramidal, do depend upon n. Molecules with 8n electrons will be linear (planar) and those with one more electron will be bent (pyramidal). This result comes from the observation that bending only occurs when all of the ligand lone pair orbitals are

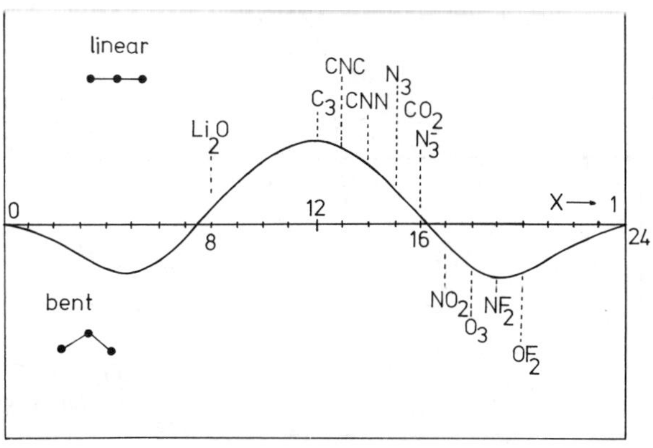

Fig. 41. A $\Delta E_4(x)$ curve on which are superimposed the electron counts for a selection of first row triatomic molecules

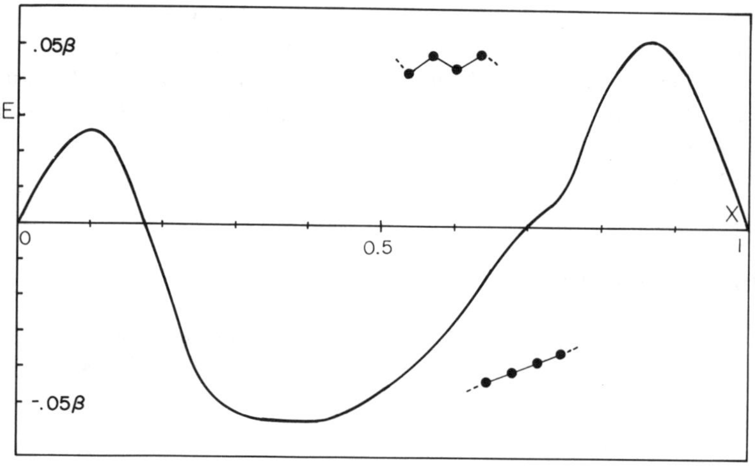

Fig. 42. Computed energy difference curve for a linear versus bent chain of carbon atoms

full and the next electron must occupy a stereochemically active central atom orbital. Thus the fractional orbital occupancy defined by the Walsh numbers, n/(n + 1) depends upon n, the coordination number. So the location of the fourth moment crossing point does not separate the 24 electron molecule BF_3 (planar), from the 25 electron CF_3 (pyramidal) and 26 electron NF_3 (pyramidal) species. The detailed agreement with experiment found in Fig. 41 is then almost certainly fortuitous.

Figure 42 shows the energy difference curve computed[11)] for linear and bent mononuclear chains with parameters appropriate for carbon. The structures of elemental sulfur, selenium and tellurium, the only examples known of elemental chains, with x = 0.75 are in accord with the computed result. These consist of spirals of atoms, the coordination about each atom being nonlinear. Notice that the curve is indeed of the

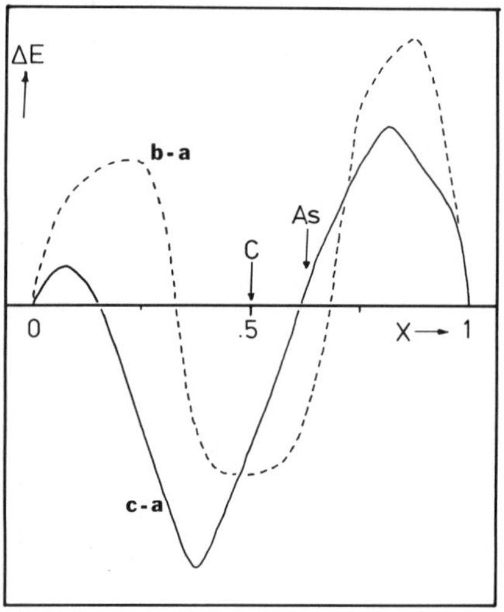

Fig. 43. Computed energy difference curves between the graphite structure and the arsenic and T structures of Fig. 44

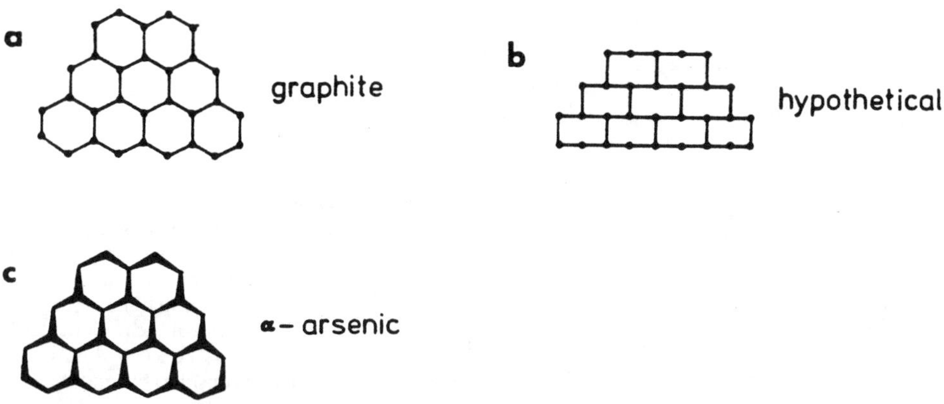

Fig. 44 a–c. Three, three coordinate structures, **a)** graphite, **b)** Structure (hypothetical) **c)** α-bismuth

form expected for a fourth moment problem, but that the crossing points are not located at exactly the same positions as in Fig. 6. Figure 43 shows computed energy difference curves for two other pairs of structures, distinguished by bond angle changes and associated this time with two-dimensional lattices (Fig. 44). The first, the graphite to arsenic distortion, is experimentally known and occurs between $x = 0.5$ (graphite) and $x = 0.625$ (group 15 elements). The other is not known. Notice that although the number of crossings associated with each curve is the same (they are both fourth moment problems) the crossing points are again system dependent. So although the general trends are in accord with the moments treatment, we need to revert to orbital arguments (with molecules especially) to locate the crossing points of the curves. We saw an exactly analogous problem in Sect. III, in the analysis of the *cis/trans* isomerism of dioxo compounds and in the discussion concerning the Jahn-Teller effect, but made little comment about it at that time. In Ref. 10 we dicuss in general some of the problems involved in locating the crossing points of our curves.

Although from **66** it is clearly the fourth moment which initially distinguishes linear and bent geometries the importance of higher moments must not be neglected. Indeed as π bonding between main group atoms decreases in importance relative to σ bonding (on going down a column of the Periodic Table) it turns out[11] that it is the fifth moment which controls the bending problem. Figure 45 shows the qualitative result expected and Fig. 46 shows the energy difference curve computed[11] for two different forms of elemental selenium and tellurium (Fig. 47). The tetragonal (hypothetical) form is one where the bond angles are 180° and the α-form the experimental geometry with nonlinear coordination at the chalcogen. As in Fig. 43 the experimental geometry for these chalcogens is the one predicted by calculation but the energy difference curve does have the shape expected from a fifth moment problem. The implications of this result are quite far-reaching. It suggests that the Walsh numbers are dependent upon the row of the Periodic Table which contains the central atom. To test this prediction we performed high quality ab initio molecular orbital calculations on N_3^- and its (unknown) second row analog P_3^-. The results[45] are shown in qualitative terms in **67.** The linear and bent forms of P_3^- are extremely close in energy, whereas for azide the linear structure is overwhelmingly

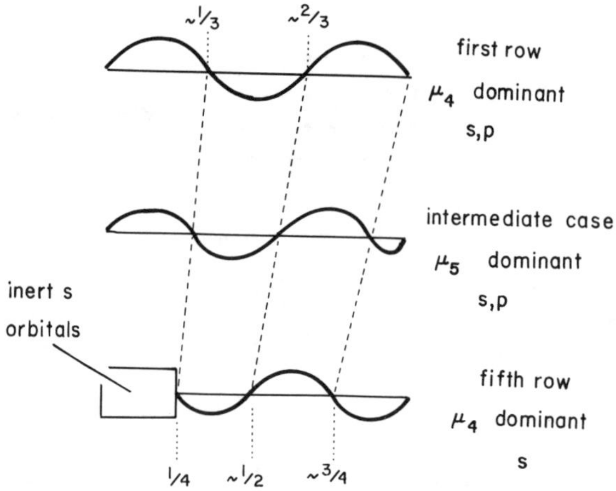

Fig. 45. Dependence of the form of the bending energy difference curve on row of the periodic table. For first row atoms, s and p orbitals are of equal importance in bonding. By the time the bottom of the periodic table is reached the valence s orbitals have contracted considerably and the picture is dominated by p orbital interaction

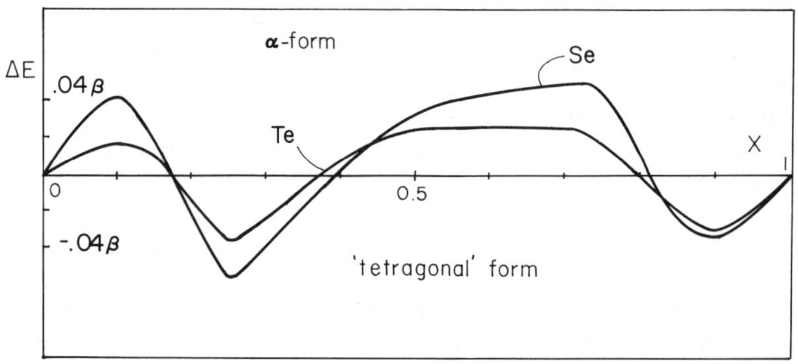

Fig. 46. Computed energy difference curves for the two forms of selenium and tellurium (one hypothetical) shown in Fig. 47

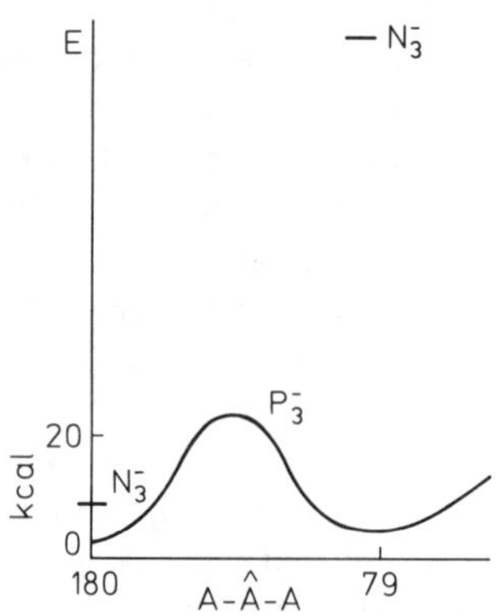

67

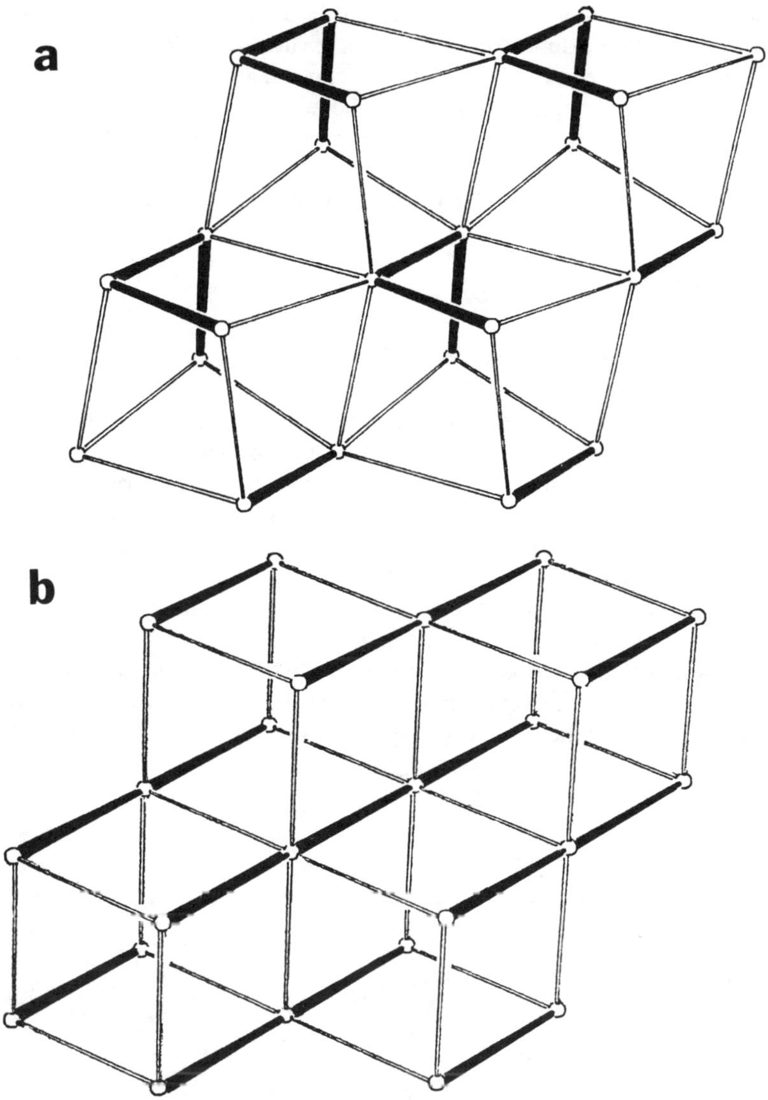

Fig. 47. **a)** Observed (α) and **b)** hypothetical (tetragonal) forms of selenium and tellurium

stabilized. The crossing point between linear and bent molecules may lie even further to lower x.

Computations[46] on the molecule Si_3(x = 0.5) also find it to be bent, stabilized relative to its linear analog by about 83.7 kJ/mol (20 kcals/mol). Such a qualitative result is easy to see using the moments approach, but the actual details of the location of the crossing points is more of a problem and may require in general the use of theory of a higher level than that of the simple or extended Hückel methods.

The results for P_3^- and Si_3 are interesting in other, more chemical ways. P_3^- is itself, energetically unfavorable at the equilateral triangular geometry since here there is a

Jahn-Teller instability. However the bent geometry computed is not far away from the triangle, found extensively in phosphorus chemistry. The geometries of P_4, P_7^{3-} and P_4S_3 present just three examples. Inclusion of σ and π bonding with the correct weights may in fact favor the formation of three membered rings for this element. The bent geometry for Si_3 may signal the preference of bent geometries for Si_4 molecules containing a central three coordinate atom. Such a result would help rationalize the nonexistence of a graphitic analog for elemental silicon, for example.

C. The Coordination Number Problem

One electron models are notoriously inaccurate when called upon to view the question of coordination number in molecules and solids. Figure 47 shows a typical result which may readily be phrased in terms of moments. It shows a calculated (Hückel) energy difference curve[11)] between bcc and simple cubic arrangements for the transition elements as a function of band filling. Notice that the simple cubic structure is (incorrectly) predicted to be more stable than bcc at all electron counts. (The bcc-fcc curve is shown for comparison). With many loops of length four in the simple cubic arrangement we expect to see a fourth moment difference curve for this plot. There is in fact a μ_4 modulation on an overwhelmingly dominant second moment curve. A much smaller effect of this type was discussed in Sect. II.H., in the comparison of boron and carbon structures. The second moment is simply given by the sum over all linkages around a given atom of the squares of the relevant interaction integrals. This becomes the sum over all the central atom orbitals i and all the orbitals j lying on adjacent atoms. In many ways then the value of μ_2

$$\mu_2 = \sum_{i,j} \mathbf{H}_{ij}^2 \quad (26)$$

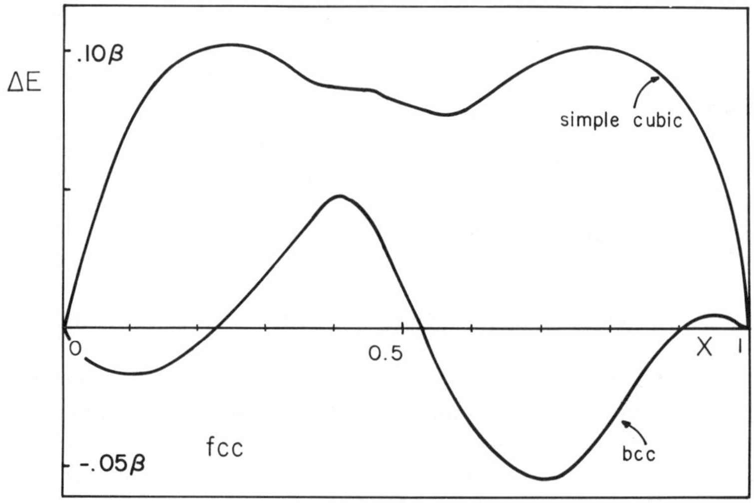

Fig. 48. Computed energy difference curve between simple cubic and fcc structures for the transition metals. The fcc/bcc curve is shown for comparison

is a good measure of the coordination number or coordination strength. The computations summarized in Fig. 48 were performed at constant density. Since there are six nearest neighbors in the simple cubic structure but fourteen in the bcc structure, the interatomic separation is considerably smaller in the former structure and the values of $\mathbf{H}_{ij}$ substantially larger. This results in a much larger μ_2 for the simple cubic compared to bcc structure in spite of there being a smaller number of contact in the former. In many places in this article, therefore, the second moments of the two structures being compared have been taken quite arbitrarily to be equal in order to avoid such coordination number problems.

However in structure where both variants have the same second moment because of the presence of the same number of close contacts, the moments approach does make some predictions as to the likely arrangement. For example the two arrangements shown in **68** and **69** have the same second and third moments but differ at the fourth moment. It

68

69

is easy to see that **69** has the larger fourth moment because of the presence of a larger number of walks of the type shown in **70.** In other words the terminal and inner atom orbitals of **68** are metriotic but the orbitals of the central, three coordinate atom, and one

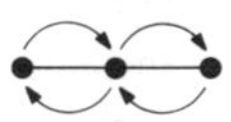

70

of the terminal atoms in **69** are busy and idle respectively. This result is considerably upset when the atoms have different electronegativities since now the fifth moment differences may be very important. The experimental observation is that the structure **68** is found when the atoms have similar electronegativities (O_2F_2, S_4^{2-}) but the structure **69** becomes more favorable when the electronegativity difference is larger (S_2F_2 is found in both forms, for example). Along similar lines recall the stabilization of square cyclobutadiene by suitable coloring and the stabilization of the $4 8^2$ net of **4** by coloring by B and C. Analogously the "tetragonal" carbon structure, not found for the elemental material by virtue of the proliferation of four-numbered rings is found in a colored version as β-BeO.

V. Conclusions

One point which it is important to remember concerning the approach described in these pages is that it is based upon the ideas of simple Hückel theory. The type of problem therefore which may be viewed in terms of moments is therefore at present restricted to those for which such a topologically based theory is applicable. We have already disqualified most coordination number problems, for example. The numerical results too which such one-electron models give are often open to doubt. Although the general trends concerning the stability of one structure over another, as a function of electron count are fine, the actual energy differences which are computed should be interpreted with care. The approach is clearly at its best when viewing, at low resolution, structural problems which are controlled by variation of electron count. It is much less applicable to the study of structural differences within a column of the Periodic Table (phosphorus chemistry versus arsenic chemistry for example) since these are isoelectric problems, although some of the qualitative suggestions concerning first/second row differences perhaps can be quantified in some way as described in Sec. IV.B. It would not be of much value either, we think, in the study of the inumerable structures adopted by silicate materials, for example. Here the structural problem is probably dominated by a soft bending potential around the SiOSi angle, and the strong electrostatic forces involved in the attachment of the electropositive ions to the cage walls. Neither do we feel it would have any special advantage over more conventional orbital methods in understanding why one particular system has a larger Jahn-Teller, or Peierls distortion compared to another isoelectronic example.

The topological component of the moments approach via the walks of **1** is a very strong one indeed. The ideas concerning the stability of types of σ framework as a function of electron count are similar in nature to Dewar's concept[47] of σ-aromaticity. We have noted in Sect. N their relationship too to the Hückel-Möbius concept. In the area of organometallic chemistry Mingos' ideas[48] concerning the topological control of energy level patterns and hence structural stability as a function of electron count may well turnout to be analogously related. We show elsewhere[49] how the use of some of the basic ideas behind the moments method may be used to provide new insight into the observation of superdegenerate orbitals in molecules. These are orbitals with a degeneracy which is higher than that allowed either by the molecular point symmetry, or the permutational symmetry. They arise via the connectivity of the atoms in the molecule, the way one atom "sees" another via a walk along the bonds of the molecule. The topological control of electronic and geometrical structure is indeed a strong one. It will be interesting to see whether these ideas can be merged with other recent topological views of solid state structure such as that associated with the generation of minimal surfaces[50].

Acknowledgements. Our work at Chicago has been funded by the National Science Foundation and by the donors of the Petroleum Research Fund, Administered by the American Chemical Society. This article was written during the tenure of a Wilsmore Fellowship at the University of Melbourne, Australia. I would like to thank the Department of Chemistry for their hospitality during this time and Dr. Michael Hitchman of the University of Tasmania for some stimulating discussions concerning the Jahn-Teller theorem. Special thanks are due to Mrs. M. Hyde who suggested the use of the term metriotic to describe orbitals which are neither busy nor idle, and to Stephen Lee whose efforts were largely responsible in making the moments method a tool chemists could use.

VI. References

1. For example, Albright, T. A., Burdett, J. K., Whangbo, M.-H.: Orbital Interactions in Chemistry, Wiley, 1985
2. For example, Burdett, J. K.: Prog. Solid State Chem. *15,* 173 (1984)
3. Cyrot-Lackmann, F.: Thèse, Orsay, 1968
4. Ducastelle, F., Cyrot-Lackmann, F.: J. Phys. Chem. Solids *31,* 1295 (1970); *32,* 285 (1971)
5. Cyrot-Lackmann, F.: J. Phys. Chem. Solids *29,* 1235 (1968)
6. Cyrot-Lackmann, F.: Surf. Sci. *15,* 535 (1969)
7. Gaspard, J.-P., Cyrot-Lackmann, F.: J. Phys. *C6,* 3077 (1973)
8. Lambin, Ph., Gaspard, J.-P.: Phys. Rev. *B26,* 4356 (1982)
9. The approach has much akin with the recursion method of solid state physics. For a recent, quite comprehensive work see: The recursion method and its applications, (Pettifor, D. G., Weaire, D. L., Eds.) Springer Series in Solid State Science *58* (1985)
10. Burdett, J. K., Lee, S.: J. Am. Chem. Soc. *107,* 3050 (1985)
11. Burdett, J. K., Lee, S.: J. Am. Chem. Soc. *107,* 3063 (1985)
12. Burdett, J. K., Lee, S., McLarnan, T. J.: J. Am. Chem. Soc. *107,* 3083 (1985)
13. Burdett, J. K., Lee, S.: J. Solid State Chem. *56,* 211 (1984)
14. Burdett, J. K., Lee, S., Sha, W. C.: Croat. Chim. Acta *57,* 1193 (1984)
15. Burdett, J. K. in: Molecular Structures and Energetics, (Greenberg, A., Liebman, J. F., Eds.), Verlag Chemie, 1987
16. There is a nice discussion of this problem in: Jones, D. E. H.: The Inventions of Daedalus, Freeman, 1982. The figure of 260,000 carbon atoms is the number calculated for optimal stability. A sixty atom cluster of this type may have been made (see J. Amer. Chem. Soc. *107,* 7779 (1985)
17. Krogh-Jesperson, K., Cremer, D., Poppinger, D., Pople, J. A., Schleyer, P. v. R., Chandrasekhar, J.: J. Am. Chem. Soc. *101,* 4843 (1979)
18. See, for example, Burdett, J. K.: Molecular Shapes, Wiley, 1980
19. For example $M(O)_2X_4$ ($X = CN^-$, CO; M = Mo, w). Lippard, S. J., Russ, B. J.: Inorg. Chem. *6,* 1943 (1967); Day, V. W., Hoard, J. L.: J. Am. Chem. Soc. *90,* 3374 (1968); Crayston, J. A., Almond, M. J., Downs, A. J., Poliakoff, M., Turner, J. J.: Inorg. Chem. *23,* 3051 (1984)
20. Burdett, J. K.: Inorg. Chem. *24,* 2244 (1985)
21. Burdett, J. K., Albright, T. A.: Inorg. Chem. *18,* 2112 (1979)

21a. Mingos, D. M. P.: J. Organomet Chem. *179,* C29 (1979)

22. Reinen, D., Friebel, C.: Struct. Bond. *37,* 1 (1979)
23. Brown, I. D.: Acta Cryst. *B33,* 1305 (1977)
24. Burdett, J. K.: Inorg. Chem. *20,* 1959 (1981)
25. Gazo, J., Bersuker, I. B., Garaj, J., Kabesova, M., Kohout, J., Langfelderova, M., Melnik, M., Serator, M., Valach, V.: Coord. Chem. Rev. *19,* 253 (1976)
26. Burdett, J. K.: Inorg. Chem. *14,* 931 (1975)
27. Shaik, S. S., Hiberty, P. C.: J. Am. Chem. Soc. *107,* 3089 (1985)
28. Burdett, J. K., Lee, S.: J. Am. Chem. Soc. *105,* 1079 (1983)
29. Burdett, J. K., Canadell, E., Hughbanks, T.: J. Am. Chem. Soc. *108,* 3971 (1986)
30. Widera, A., Schafer, H.: Z. Naturf. *B34,* 1769 (1979)
31. Bieber, A., Ducastelle, F., Gautier, F., Treglia, G., Turchi, P.: Sol. State Comm. *45,* 585 (1983)
32. Burdett, J. K., Lawrence, N. J., Turner, J. J.: Inorg. Chem. *23,* 2419 (1984)
33. Oka, T.: Phys. Rev. Lett. *45,* 531 (1980)
34. Upmaris, R. K., Gadd, G. E., Poliakoff, M., Simpson, M. B., Turner, J. J., Whyman, R., Simpson, A. F.: J. Chem. Soc., Chem. Comm. *27* (1985); Church, S. P., Grevels, F.-W., Herman, H., Schaffner, K.: J. Chem. Soc., Chem. Comm. *30* (1985)
35. Kubas, G. J., Ryan, R. R., Swanson, B. I., Vergami, P. J., Wasserman, H. J.: J. Am. Chem. Soc. *106,* 451 (1984)
36. Radom, L., Poppinger, D., Haddon, R. C., in: Carbonium Ions, Vol. V, Chapter 38, (Olah, G. A., Schleyer, P. v. R., Eds.) Wiley, 1976

37. Collins, J. B., Schleyer, P. v. R., Binkley, J. S., Pople, J. A., Radom, L.: J. Am. Chem. Soc. *98,* 3436 (1976)
38. Woodward, R. B., Hoffmann, R.: J. Am. Chem. Soc. *87,* 395, 2046, 2511, 4389 (1985)
39. Woodward, R. B., Hoffmann, R.: Accts. Chem. Res. *1,* 17 (1968)
40. Zimmerman, H. E.: Accts. Chem. Res. *4,* 272 (1971)
41. See, for example, Yates, K.: Hückel Molecular Orbital Theory, Academic Press, 1978
42. See, for example, Pettifor, D. G.: Calphad *1,* 305 (1977)
43. Streitweiser, A.: Molecular Orbital Theory for Organic Chemists, Wiley, 1961
44. Ault, B. S., Andrews, L.: Inorg. Chem. *16,* 2024 (1977)
45. Burdett, J. K., Marsden, C. J.: (submitted for publication)
46. Diercksen, G. H. F., Grüner, N. E., Oddershede, J., Sabin, J. R.: Chem. Phys. Lett. *117,* 29 (1985)
47. Dewar, M. J. S.: Bull. Chim. Soc. Belgae *88,* 957 (1979)
48. Mingos, D. M. P.: J. Chem. Soc. (Dalton) 20, 26, 31 (1977)
49. Burdett, J. K., Lee, S., Sha, W. C.: Nouv. J. Chimie *90,* 757 (1985)
50. Andersson, S., Hyde, S. T., von Schnering, H. G.: Z. Kristallog. *168, 1* (1984); Hyde, S. T., Andersson, S., Ericsson, B., Larsson, K.: Z. Kristallog. *168,* 213 (1984)

Electronic Structure of Intermetallic B32 Type Zintl Phases

Peter C. Schmidt

Institut für Physikalische Chemie, Physikalische Chemie III, Technische Hochschule Darmstadt, Petersenstr. 20, D-6100 Darmstadt, Federal Republic of Germany

The electronic properties of binary and ternary intermetallic Zintl phases crystallizing in the NaTl type of structure are investigated and reported. The crystal and defect structure, the electronic band structure and density of states, the bonding mechanisms and the charge transfer are discussed. Comparison is made between the electronic states in the B2 and the B32 types of structure. Furthermore, based upon theoretical studies of the electronic valence states the optical properties (imaginary part of the dielectric constant and reflectivity) and the magnetic properties (susceptibility and Knight shift) are considered.

A. Introduction 92

B. Crystal Structure 93

C. Electronic Structure 98
C. I. Methods 98
C. II. Band Structure and Density of States 103
C. III. Fermi Energy 105
C. IV. Charge Distribution 106
C. V. Charge Transfer 112
C. VI. Ternary Systems 114

D. Comparison of the Electronic States in the B2 and B32 Structure 116
D. I. Comparison of the Band Structures 117
D. II. Energy Differences 118

E. Optical and Magnetic Properties 121
E. I. Optical Properties 121
E. II. Magnetic Susceptibility 123
E. III. Knight Shift 126

F. Conclusions 130

G. References 131

Structure and Bonding 65

A. Introduction

At present a large variety of solid compounds are called Zintl phases[1]. The name Zintl phase was introduced by Laves[2]. According to Laves, Zintl phases are those intermetallic compounds which crystallize in typical "non-metal" crystal structures. For these compounds one expects an ionic contribution to the chemical bond. This definition has been extended[3] to a large number of solid compounds formed by alkali or alkaline earth metals with metallic or semimetallic elements of the fourth, fifth and partly third group of the Periodic Table for which common structural and bonding properties have been found. The crystal structures and chemical properties of these compounds have been studied extensively[3].

The present work will only consider the electronic properties of those Zintl phases which crystallize in the B32 structure. Binary B32 Zintl phases are formed by the AB^{IIIa} compounds LiAl, LiGa, LiIn and NaIn, by the prototype NaTl, and by the AB^{IIb} compounds LiZn and LiCd. Ternary systems, which crystallize in this structure, are found for alloys $AB_{1-x}C_x$, for which at least one boundary compound AB or AC crystallizes in the NaTl type of structure.

Since the X-ray diffraction studies of Zintl et al.[4,5], these members of the family of intermetallic compounds have been of special interest because some of their chemical properties are unusual for intermetallic phases. Many experimental investigations have been reported for binary and ternary B32 type compounds. Besides the crystal structure[4–18], the thermodynamic behavior[19,20], electrical conductivity[20–24], magnetic susceptibility[25,26], NMR data[17,18,27–32], elastic constants[33,34] and optical properties[35,36] have been studied. Additionally for LiAl electrochemical investigations have been performed in view of the recent interest in fast ionic conductors[37–39].

Based upon the experimental data it was deduced that the chemical bonding in Zintl phases should be a mixture of covalent, ionic and metallic contributions[40–44]. As metal-like systems the B32-type compounds $A_{1-x}B_x$ possess a distinct phase width in the range of approximately $0.45 \leq x \leq 0.55$[45]. Furthermore one finds a metallic conductivity in these systems and as in metals the electrical resistivity increases with temperature[23].

On the other hand from the thermic and elastic behaviour of these phases one could conclude that strong covalent bond contributions are present in B32-type phases[46]. The magnetic measurements show deviations from the typical paramagnetism of metals[25,46]. The chemical shifts in the NMR signals (Knight shift) of the AB^{IIIa} compounds are much smaller than in the pure parent metals[47], and at least for NaTl the valence electrons show a large diamagnetic contribution to the magnetic susceptibility[25]. Finally it has been argued that an ionic bonding component might result from a charge transfer from the alkali to the non-alkali atoms. A charge transfer is indicated by the value of the measured Knight shift K_s at the alkali nuclei. $K_s(\underline{A}B)$ is almost zero[27,32]; that is, one gets the same resonance frequency as in the reference ionic solution.

Indeed, also the differences in the electronegativities of the A and B in AB in relation to the arrangement of the atoms in the lattice suggest that a charge transfer might exist[48]. The arrangement of the A and B atoms in the B32 structure is unusual for AB alloys. Normally all nearest neighbours of an atom in the lattice are of the same kind whereas in the NaTl structure each atom Na(Tl) possesses four equal Na(Tl) and four different Tl(Na) neighbours. Each atom forms a diamond like sublattice. The diamond type of structure is usually found for tetravalent atoms. In the valence state picture the covalent bonds in these crystals are formed by saturated sp^3-like hybrids, and a semi-conducting behaviour results. In NaTl the trivalent Tl might receive the 4th electron from Na to form fully occupied valence states (8-N rule, see[1, 6, 49–53]). However, the mixture of metallic and covalent properties in Zintl phases mentioned above does not support a charge distribution Na^+Tl^-. Moreover, saturated sp^3-hybrids cannot be formed by the binary AB^{IIb}[5] and the ternary B32-type compounds.

These interesting and somewhat unusual properties associated with the B32 type Zintl phases have generated theoretical interest in this area[35, 54–64]. In the present work theoretical investigations of B32 type intermetallic phases will be reported. Furthermore, the optical and magnetic properties found experimentally are interpreted on the basis of the electronic structure of the B32-type compounds gained from band structure calculations[17, 18, 35, 61, 63].

In Sect. B, structural properties such as phase widths and bonding distances are considered. The bond distances in B32 type Zintl phases have been discussed extensively in the literature because it is expected from the arrangement of the two joined diamond sublattices that both kinds of atoms should have the same size[43, 55].

In Sect. C, the band structure data based on self-consistent relativistic augmented-plane-wave calculations performed by the author[61–63] are presented. Besides the electronic bands and the densities of states, the nature of the chemical bond is discussed. In Sect. D the electronic states in Zintl phases are compared with those having the B2 type of structure. As shown in Sect. B the B2 structure is closely related to the B32 structure. For intermetallic compounds the B2 structure seems to be the more "natural" because in this lattice all nearest neighbours of an atom A are B atoms. The reason why the compounds mentioned above crystallize in the B32 structure whereas similar compounds like LiTl and KTl form B2 phases has been frequently discussed in the literature[43, 54–56, 61, 64, 65].

Finally, in Sect. E the optical and magnetic properties are considered. It is found experimentally[11, 35] that some Zintl phases are colored and in ternary systems the color changes continuously as a function of the composition. This change can be correlated to a shift in a maximum of the imaginary part ε_2 of the dielectric constant ε, and ε_2 can be interpreted by electronic interband transitions[35]. The magnetic susceptibility and Knight shift are discussed on the basis of spin polarized band structure calculations[61, 63]. Spin and orbital contributions are also considered.

B. Crystal Structure

The arrangement of the atoms in the NaTl structure and its relation to some other simple cubic structures is illustrated in Fig. 1 and Table 1. In Fig. 1a a unit cell is shown with four different lattice sites A to D. The A2 structure results if all four sites are occupied by the same kind of atoms or statistically by different atoms. For AB compounds the CsCl structure results for the configurations A_A, A_B, B_C and B_D (cf. Fig. 1b). The NaTl structure follows from A_A, B_B, A_C and B_D as displayed in Fig. 1c. For ternary Zintl phases A_2BC one finds that either the B and C atoms are statistically distributed on the B and D sites given in Fig. 1a (B32A structure) or they are ordered and the XA type of structure results[11, 15] (cf. Table 1).

Table 1. Cubic crystal structures on the basis of the lattice sites displayed in Fig. 1a (A_B = atom A on lattice site B)

Structure	Prototype	Positions of the atoms				Space group
A2	W	A_A	A_B	A_C	A_D	O_h^9-Im3m
B2	CsCl	A_A	A_B	B_C	B_D	O_h^1-Pm3m
B32	NaTl	A_A	B_B	A_C	B_D	O_h^7-Fd3m
XA	Li_2MgSn	C_A	A_B	A_C	B_D	T_d^2-F43m
A4	C (diamond)	A_A	V_B	A_C	V_D	O_h^7-Fd3m

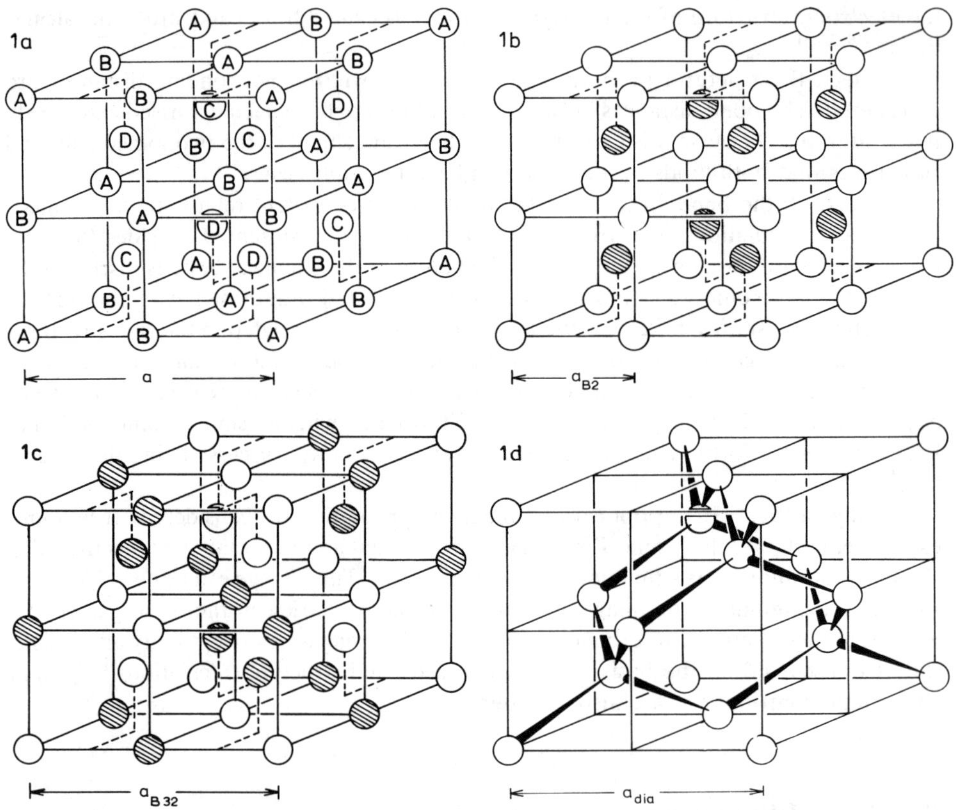

Fig. 1a–d. Arrangements of the atoms in some cubic structures, **a)** Lattice sites A to D in superstructures of the bcc lattice, see Table 1, **b)** Unit cells of the B2 type of structure, **c)** Unit cell of the B32 type of structure, **d)** Unit cell of the diamond type of structure

From Fig. 1c one notices that in the NaTl structure both the Na and Tl atoms form diamond-like sublattices (Fig. 1d). The diamond structure arises from the NaTl structure by leaving vacancies V in one sublattice (Table 1). Therefore in the B32 structure (Fig. 1c) the bond lengths d(A–A), d(A–B) and d(B–B) are equal and the atoms A and B must have equal sizes to get a spacefilling distribution[69]. It follows that the atomic radii in B32 type compounds r_{B32} should be half of the distance to the nearest neighbours. In Table 2 the lattice constants for the binary Zintl phases are listed. The resulting atomic radii r_{B32} are shown in Table 3. For comparison the Goldschmidt atomic radii r_{atom}[70] in metallic systems are given too. As can be seen from the last column of Table 3 r_{B32} and r_{atom} differ by less than or about 10%. The ratios r_{B32}/r_{atom} are smaller for the sodium compounds than for the lithium compounds. In those cases where r_{B32} and r_{atom} are not equal, the separations of the larger atoms, say B, B–B and A–B, must either be compressed[55], or a charge transfer to the more electronegative non-alkali element or covalent bonds within the non-alkali sublattices reduce the effective atomic radii in Zintl phases compared to pure metallic bonding. Therefore, the ionic radii r_{ion} of the alkali atoms and the covalent bond radii r_{cov} of the non-alkali atoms are also listed in Table 3. Some aspects of the role of the sizes of the atoms on the stability of the structures are discussed in Sects. C.VI and D.

As already mentioned in the introduction binary B32 type compounds A_xB_{1-x} are found within a range of composition of approximately $0.45 \lesssim x \lesssim 0.55$[45]. For LiAl the

Table 2. Lattice constants a for binary B32 type compounds, Zintl et al.[4, 5]

	a/Å
LiZn	6.209
LiCd	6.687
LiAl	6.360
LiGa	6.195
LiIn	6.786
NaIn	7.297
NaTl	7.473

Table 3. Atomic radii (in Å) of the atoms which form B32 type compounds. r_{B32} is the half nearest neighbour distance in Zintl phases AB, r_{atom} is the atomic radius according to Goldschmidt[67–69] (C.N.: 8), r_{ion} is the ionic and r_{cov} the covalent radius for tetrahedral coordination (C.N.: 4) according to Pauling[70, 71]

Atom	Zintl phase	r_{B32}	r_{atom}	r_{ion}	r_{cov}	r_{B32}/r_{atom}
Li	LiZn	1.34	1.51	0.78		0.89
	LiCd	1.45				0.96
	LiAl	1.38				0.91
	LiGa	1.34				0.89
	LiIn	1.47				0.97
Na	NaIn	1.58	1.86	0.96		0.85
	NaTl	1.62				0.87
Zn	LiZn	1.34	1.33		1.31	1.01
Cd	LiCd	1.45	1.48		1.48	0.98
Al	LiAl	1.38	1.36		1.26	1.01
Ga	LiGa	1.34	1.39		1.26	0.96
In	LiIn	1.47	1.52		1.44	0.97
	NaIn	1.58				1.03
Tl	NaTl	1.62	1.66		1.47	0.98

non-stoichiometric structure has been investigated by X-ray, density and NMR measurements[16, 72]. Comparing the results of these measurements with the possible types of defect structures it follows that two defect mechanisms should be dominant: vacancies V_{Li} at the Li sites and substitutional Li_{Al} in the Al sublattice. Kishio and Brittain[16] have calculated the concentration of the vacancies $[V_{Li}]$ from the difference between the densities gained from X-ray and pyknometric density measurements respectively:

$$[V_{Li}] = \frac{d_{X\text{-ray}} - d_{pyk.}}{d_{X\text{-ray}}} \tag{1}$$

The concentration of the substitutional Li_{Al} is determined from the equation

$$[Li_{Al}] = (1 - x)[V_{Li}] + x - 0.5 \tag{2}$$

These equations are used to derive a defect structure for LiAl as summarized in Table 4. The V_{Li} defects are dominant for low Li concentrations whereas with increasing x Li_{Al} defects become more important.

It follows from NMR investigations that defects are also present in the other AB^{IIIa} B32 type Zintl phases, and the defect structure seems to be similar to the one of LiAl[16, 63, 72]. For the AB^{IIb} compounds, however, no defects could be detected from NMR measurements.

The interplay of the two defects, V_{Li} and Li_{Al}, has an interesting effect on the number of valence electrons in the crystals Li_xAl_{1-x}. In the present work the valence electron concentration c_{VE} is defined by the number of valence electrons per lattice site. For the stoichiometric phase AB with the ideal lattice c_{VE} is equal to the electron-to-atom ratio e/a (1.5 for AB^{IIb} and 2.0 for AB^{IIIa}). In the last column of Table 4 c_{VE} is given as a function of x for the Li_xAl_{1-x} phases:

$$c_{VE} = [Li_{Li}] + [Li_{Al}] + 3\,[Al_{Al}]\ . \tag{3}$$

One recognizes that the interplay between V_{Li} and Li_{Al} leaves the valence electron concentration nearly constant. Furthermore, c_{VE} is always smaller than the ideal value of 2. In view of the electronic structure of the B32-type compounds the values $c_{VE} < 2$ have

Table 4. Lattice parameters a, density $d_{pyk.}$ and calculated defect structure of B32 type compounds Li_xAl_{1-x} from Kishio and Brittain[16]. c_{VE} is the number of valence electrons per lattice site calculated from Eqs. (1) and (2)

x	a/Å	$d_{pyk.}/(g/cm^3)$	$[V_{Li}]$	$[Li_{Al}]$	$[Li_{Li}]$	$[Al_{Al}]$	c_{VE}
0.483	6.3612	1.727	0.033	0.000	0.467	0.500	1.967
0.489	6.3639	1.725	0.026	0.002	0.474	0.498	1.970
0.494	6.3671	1.715	0.024	0.006	0.476	0.494	1.964
0.497	6.3685	1.709	0.024	0.009	0.476	0.491	1.958
0.503	6.3710	1.708	0.016	0.011	0.484	0.489	1.962
0.508	6.3743	1.702	0.012	0.014	0.484	0.486	1.956
0.519	6.3782	1.684	0.007	0.023	0.493	0.477	1.947
0.525	6.3825	1.669	0.007	0.029	0.493	0.471	1.935
0.539	6.3870	1.647	0.002	0.040	0.498	0.460	1.918

consequences for the filling of the Jones zones (or Brillouin zones), which is important for the interpretation of the magnetic properties (cf. Sect. E.III). The comparison of experimental and theoretical Knight shift data especially suggests that for all $A_{0.5}B_{0.5}$ B32 type compounds c_{VE} should be more than 0.5 to 3.5% smaller than the value for the perfect lattice ($c_{VE} = 2$)[63].

Finally ternary systems shall be considered. The structure of a large number of ternary systems of the composition $LiB_{1-x}C_x$ has been investigated[9–12, 15, 17]. For those systems, for which both boundary phases of the quasibinary intersection LiB and LiC are B32 type compounds, the B32A structure is found within the whole range of composition $0 \leq x \leq 1$. The lattice constants vary linearly as a function of x (Vegards law). For those compounds, for which either LiB or LiC does not crystallize in the B32 structure, the XA structure is often realized. For these compounds $LiB_{1-x}C_x$ Zintl phases are only detected for the range of composition, in which the valence electron concentration remains in the range of $1.5 \leq c_{VE} \lesssim 2.0$.

In Figs. 2 and 3 the changes of the lattice constants as a function of x are displayed for two examples, $LiCd_{1-x}Tl_x$ and $NaHg_{1-x}Tl_x$. For the system $NaHg_{1-x}Tl_x$ the B32 structure is found over a wide range of composition, whereas for $LiCd_{1-x}Tl_x$ a large two-phase region is found. In Sect. E the chemical shifts in these phases will be discussed.

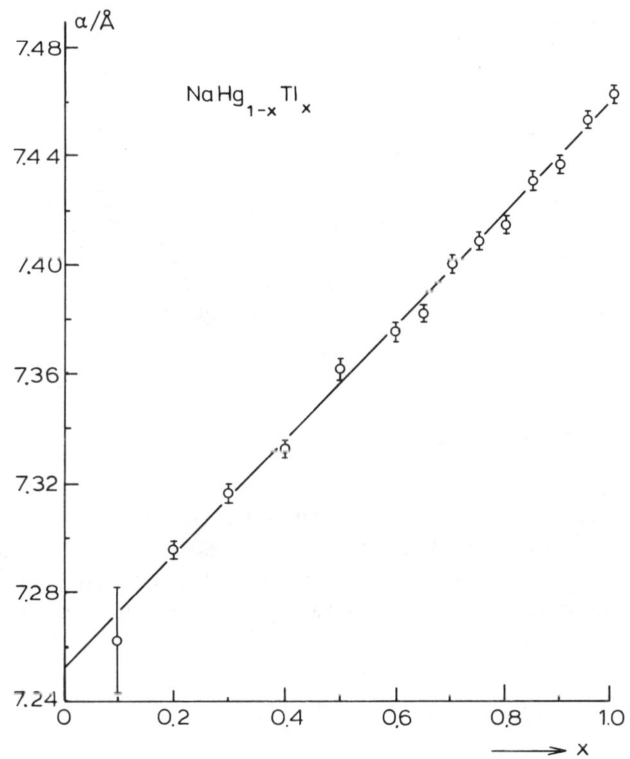

Fig. 2. Lattice constants as a function of x for B32 type compounds $NaHg_{1-x}Tl_x$, $0.1 < x \leq 1$ (Schmidt et al.[18])

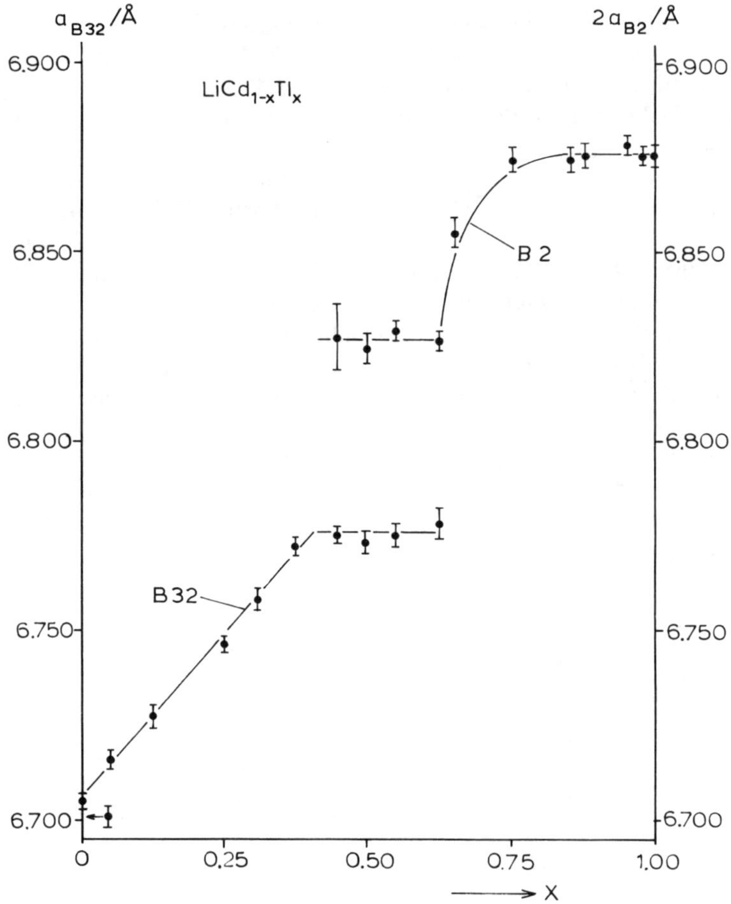

Fig. 3. Lattice constants a_{B32} and 2 a_{B2} as a function of x for B32 type and B2 type compounds $LiCd_{1-x}Tl_x$ (Baden et al.[17]). There is a two-phase region within the range of composition $0.45 < x < 0.625$

C. Electronic Structure

This section reports about theoretical investigations of the band structures and charge distributions of B32 type Zintl phases. In the first subsection the methods used for calculating the band structure data of the corresponding phases are briefly outlined. In the remaining subsections various aspects of the chemical bonding in Zintl phases are analyzed.

C.I. Methods

The one-electron wave functions ψ_i of the crystal are usually expanded in basis functions. Two types of functions have been used for investigating B32 type Zintl phases: linear combinations of atomic orbitals (LCAO)[57–60, 64] and augmented plane waves (APW)[35, 61–63]. In this section those aspects of the APW method[63, 64] which are important for understanding the calculated physical properties presented in the following subsections are explained. At the end of this section the differences between the APW approach and the simpler LCAO formalisms will be specified.

The electronic charge density $\varrho(\vec{r})$ in the crystal is calculated from the one-electron wave functions ψ_i by summing over all the occupied electronic states i:

$$\varrho(\vec{r}) = \sum_{i}^{\text{occ.}} |\psi_i(\vec{r})|^2 \tag{4}$$

In non-relativistic approaches the functions ψ_i are solutions of the one-electron Schrödinger equation:

$$(\hat{T} + V(\vec{r}))\,\psi_i(\vec{r}) = \varepsilon_i \psi_i(\vec{r}) \tag{5}$$

where $\hat{T}$ is the operator for the kinetic energy and $V(\vec{r})$ is the crystal potential "seen" by the ith electron.

As shown above several B32 type Zintl phases contain heavy atoms like Cd, In, Hg, and Tl. For theoretically investigating such phases relativistic quantum mechanics has to be used. In the relativistic version of the APW procedure (RAPW)[73, 74] the one-electron wave functions ψ_i are 4-spinors and each ψ_i is a solution to the one-electron Dirac equation[75] which is equivalent to Eq. (5). According to the Dirac theory, however, the operator $\hat{T}$ is a more complicated operator[73–75].

In the (R)APW procedure the crystal potential $V(\vec{r})$ in Eq. (5) is approximated by using the local density approach[76, 77], where $V(\vec{r})$ is first separated into three terms: the

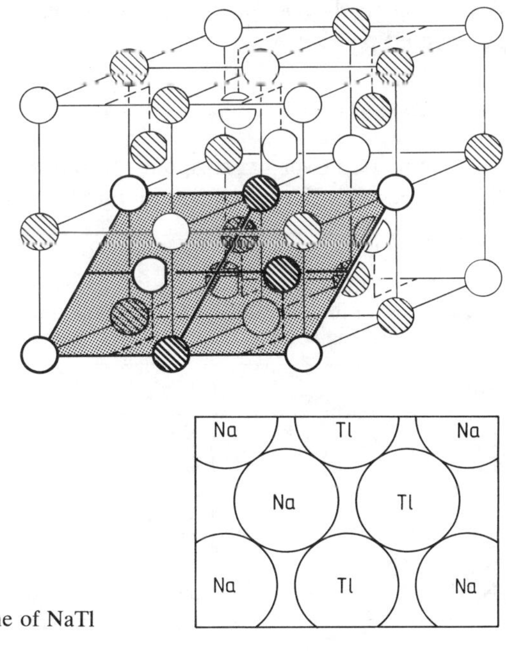

Fig. 4. Muffin-tin spheres for the (101) plane of NaTl

attraction by the nuclei, the mean Coulomb repulsion of the electrons and the rest, called the exchange-correlation term. This last term is approximated by the local density expression[76, 77]. The ansatz of Vosko et al.[78, 79] is used for the relativistic exchange-correlation potential.

Furthermore, within the (R)APW method the so called muffin-tin approach is used for calculating $V(\vec{r})$. According to this model the volume of the unit cell ω is separated into the volume ω_y of non overlapping and approximately touching atomic spheres (muffin-tin spheres, cf. Fig. 4) centred at the lattice sites y and the volume ω' between the spheres. In Table 5 the radii r_y and the volumes ω_y which are used for the RAPW calculations of Zintl phases are given. Because of the arrangement of the atoms in the crystal shown in Sect. B, the volumes ω_A and ω_B are equal in each AB compound. For comparison in Table 5 the Wigner-Seitz volumes are listed too.

According to the muffin-tin approximation the potential $V(\vec{r})$ inside the muffin-tin spheres is replaced by its spherical mean $V(r)$, and outside the atomic spheres $V(\vec{r})$ is approximated by a constant V_0. The electron-electron interaction in this region is calculated from the constant charge density ϱ_0 defined by:

$$\varrho_0 = \frac{Q'}{\omega'} \tag{6}$$

where Q' is the electronic charge in units of e^- inside the volume ω'.

The electron-electron potential and the exchange-correlation term inside the atomic spheres are determined by the spherical mean of the electronic charge density $\varrho(\vec{r})$

$$\varrho_y(r) = \frac{1}{4\pi} \int\int \varrho(\vec{r}) d\varphi d\cos\vartheta \, . \tag{7}$$

According to the (R)APW method the $\psi_i(\vec{r})$ are linear combinations of (relativistic) augmented plane waves $\varphi_{j,m}$:

$$\psi_i = \sum_{j,m} c_{j,m} \varphi_{j,m} \, . \tag{8}$$

The summation over j corresponds to the summation over the reciprocal lattice vectors $\vec{K}_j$, whereas m is the spin quantum number. Within the non-relativistic procedure m is a

Table 5. Atomic sphere radii r_y (in Å) and volumes ω_y (in Å^3) used for the APW band structure calculations for intermetallic B32-type compounds AB (Schmidt[61, 62]). ω_{WS} is the volume per lattice site (Wigner-Seitz volume)

AB	$r_A = r_B$	$\omega_A = \omega_B$	ω_{WS}
LiZn	1.301	9.22	14.96
LiCd	1.438	12.46	18.69
LiAl	1.368	10.72	16.08
LiGa	1.301	9.22	14.86
LiIn	1.438	12.46	19.52
NaIn	1.512	14.48	24.29
NaTl	1.590	16.84	26.24

good quantum number and the summation over m can be omitted. In this case the electronic state i can be described by the wave vector $\vec{k}$ of the electron, the band index n and the spin quantum number m:

$$\psi_i(\vec{r}) = \psi_{k,n,m}(\vec{r}) \; . \tag{9}$$

According to the shape of the muffin-tin potential the basis functions $\varphi_{j,m}$ in Eq. (8) are expanded inside the atomic sphere y in spin-angular functions in the relativistic case and in spherical harmonics Y_{l,m_l} in the non-relativistic case:

$$\varphi_j^{(y)}(r, \vartheta, \varphi) = \sum_{l, m_l} A_{l,m_l}^{y,j} g_l^{(y)}(r) Y_{l,m_l}(\vartheta, \varphi) \; . \tag{10}$$

Outside the atomic spheres the functions are plane waves

$$\varphi_j(\vec{r}) = \exp\{i(\vec{k} + \vec{K}_j) \cdot \vec{r}\} \; . \tag{11}$$

In Eq. (10) the coefficients $A_{l,m_l}^{y,j}$ are chosen in such a way that the values for the functions (10) and (11) become equal at the boundary of the muffin-tin sphere, and the functions $g_l^{(y)}(r)$ are the radial partial waves inside the atomic regions. These functions are exact solutions of the radial Schrödinger equation.

By interchanging the summation over the reciprocal lattice (index j) and the summation over the angular momentum in Eqs. (8) and (10), one can write:

$$\psi_{k,n}^{(y)}(\vec{r}) = \sum_{l, m_l} R_{l,m_l}^{(y)}(r) Y_{l,m_l}(\vartheta, \varphi) \; , \tag{12}$$

where the radial wave functions $R_{l,m_l}^{(y)}(r)$ depend on the $g_l^{(y)}(r)$ and on the coefficients given in Eqs. (8) and (10)[73, 74]. Using the formula in Eq. (12) the l contributions to the one-electron eigenstates can be deduced.

According to the LCAO methods used for theoretically investigating of B32 type Zintl phases, the wave functions are also expanded in spherical harmonics. However, the basis functions are properly chosen atomic orbitals with radial parts that are not exact solutions for the crystal potential used in the one-electron equation. In the first detailed charge analysis of LiAl by Zunger[57] the discrete variational energy-band 1 s, 2 s, 2 p and Al 1 s, 2 s, 2 p, 3 s, 3 p and 3 d orbitals were used for these investigations.

The investigations of Asada et al.[58, 59] and Christensen[64] were carried out with linear-muffin-tin orbitals within the atomic sphere approximation (LMTO-ASA)[80, 81]. Within the muffin-tin model suitable s, p and d basis functions (muffin-tin orbitals, MTO) are chosen. In contrast to the APW procedure the radial wave functions chosen in the linear MTO approach are not exact solutions of the radial Schrödinger (or Dirac) equation. Furthermore, in the atomic sphere approximation (ASA) the radii of the atomic spheres are of the Wigner-Seitz type (for metals the spheres have the volume of the Wigner-Seitz cell) and therefore the atomic spheres overlap. The ASA procedure is less accurate than the APW method. However, the advantage of the ASA-LMTO method is the drastic reduction of computer time compared to the APW procedure.

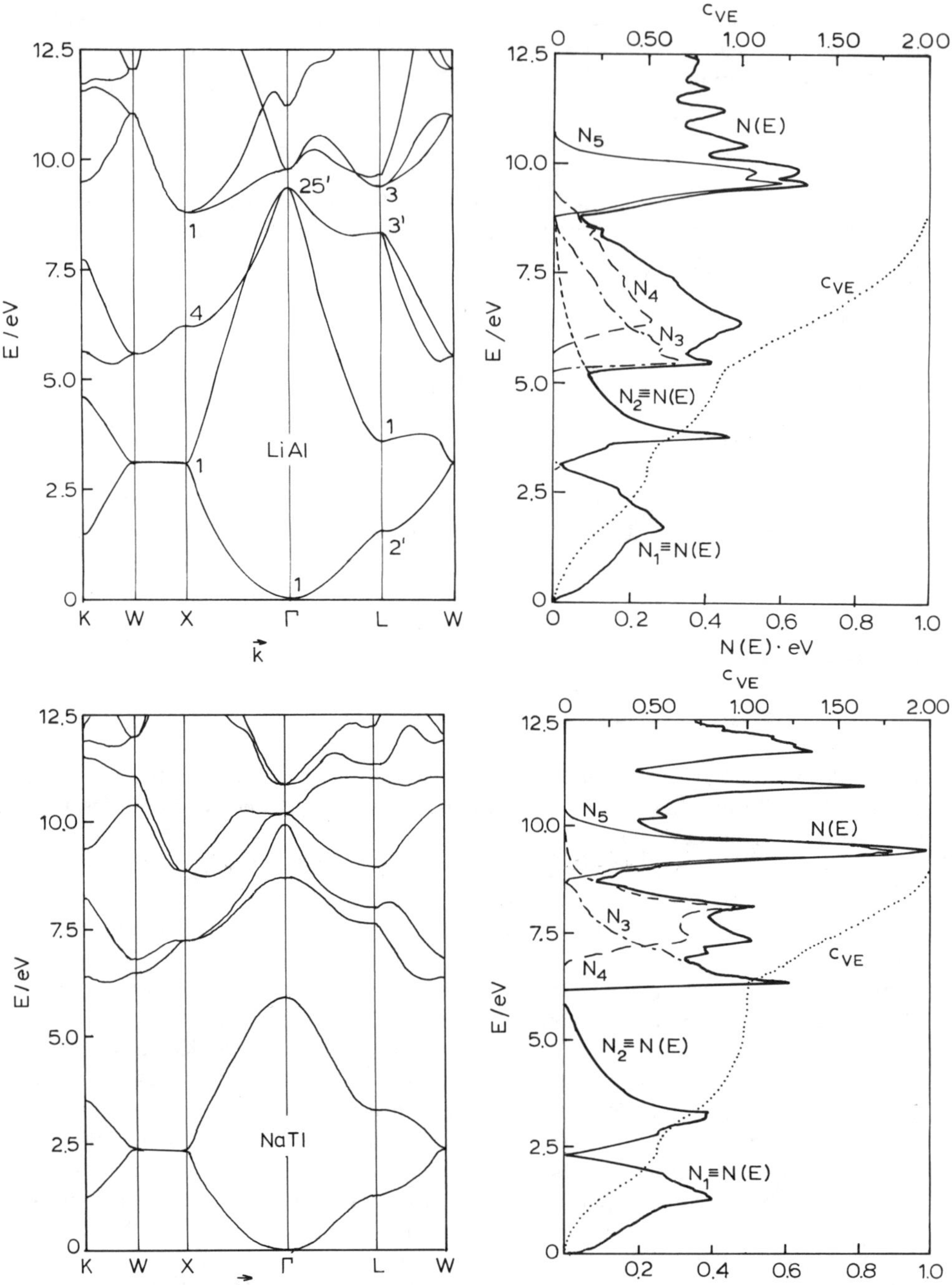

Fig. 5. *Left:* Electronic band structure of the valence states of LiAl and NaTl. *Right:* Density of states N(E) per lattice site. For the first five bands the bandwise contribution to N(E), $N_n(E)$, are also plotted. The position of the Fermi energy can be deduced from the number of occupied states per lattice site c_{VE} as a function of E. For the defect phase LiAl N_{VE} is smaller than the ideal value $c_{VE} = 2$, see Table 4 (Schmidt[61, 62])

C.II. Band Structure and Density of States

In this subsection band structures of B32 type compounds determined by the author[61, 62] using the relativistic augmented plane wave (RAPW) method are presented. For LiAl the APW results are compared with the results obtained using the LCAO variants (Zunger[57]) and Asada et al.[58]) mentioned above.

The band structures of all binary B32 type Zintl phases are similar. Figure 5 shows the results for two examples, LiAl and NaTl, respectively. The band structures have been calculated for the rhomboedric unit cell containing two formula units AB. On the left hand side of Fig. 5 the dispersion curves $E_n(\vec{k})$ for the valence electrons are plotted as a function of $\vec{k}$ for various directions in $\vec{k}$-space. On the right hand sides of Fig. 5 the densities of states for the valence bands are plotted. Besides the total density of states N(E) the density of states is displayed bandwise for the 1st to 5th band ($N_1(E)$ to $N_5(E)$). The first and second bands show only a small overlap (cf. N_1 and N_2). For NaTl these two low lying valence bands are separated by a gap from the 3rd band. The 3rd and 4th bands show a large mutual overlap. The 5th band is the first conduction band. One finds only a small overlap between the 5th band and the 3rd and 4th bands and a deep minimum in the density of states curve between $N_4(E)$ and $N_5(E)$. As discussed below the Fermi energy E_F lies in this energy range, with small values of N(E). This is one reason why certain semimetal properties are found for the AB^{IIIa} Zintl phases. If a gap existed between $N_4(E)$ and $N_5(E)$ these phases would be semi-conductors.

Figure 6 and 7 show the density of states curves for all binary B32 type Zintl phases. The shapes of the curves are very similar. The main difference between the AB^{IIb} and the AB^{IIIa} compounds is the position of the Fermi energy. This is discussed in the next subsection.

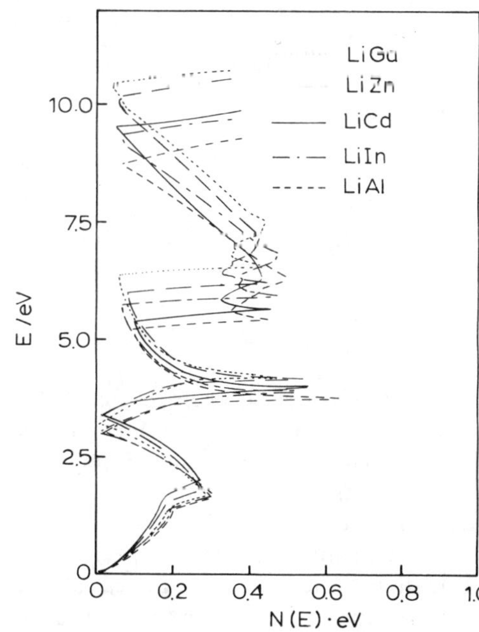

Fig. 6. Densities of states for the electronic valence states of B32 type compounds LiB as a function of the energy E

Table 6. Comparison of the energy separation (in Ry) between various levels in LiAl calculated by Schmidt[62], Zunger[57] and Asada et al.[59]

	Schmidt	Asada et al.	Zunger
Valence-valence			
$X_1 — X_4$	0.233	0.233	0.217
$L_{2'} — L_1$	0.146	0.136	0.150
$L_1 — L_{3'}$	0.341	0.367	0.339
Valence-exited			
$\Gamma_1 — \Gamma_{25'}$	0.682	0.702	0.694
$X_4 — X_1$	0.191	0.193	0.256
$L_{3'} — L_3$	0.085	0.051	0.110
$X_1 — \Gamma_{25'}$	0.040	0.062	0.013

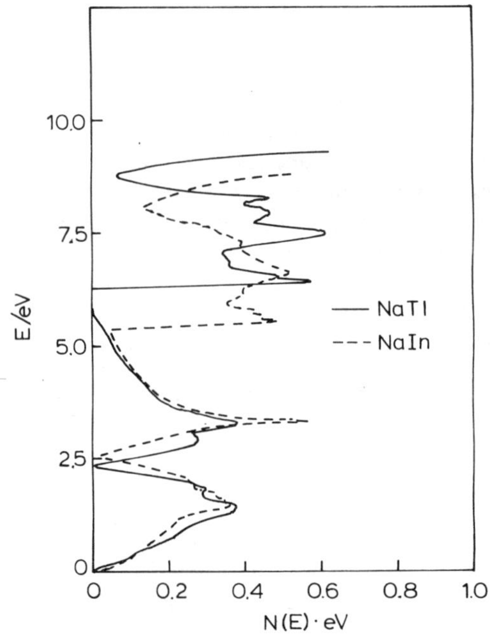

Fig. 7. Densities of states for the electronic valence states of B32 type compounds NaB as a function of the energy E

The differences between the band structures obtained from various methods are summarized in Table 6, which shows the energy separation between some levels of LiAl determined by different authors. Comparing the data given in Table 6 one sees that there is no unique trend for the energy separation ΔE. This indicates that the differences in the ΔE values are caused by differences in the hybridization of the Li and Al valence orbitals since the relative energies of the levels depend on the basis functions used for the calculations. This influences the calculated density of states and the resulting physical properties as will be seen in the next sections.

C.III. Fermi Energy

In Fig. 5 the density of states N(E) is normalized to the number of lattice sites. Then the Fermi energy E_F is defined by

$$\int_0^{E_F} N(E)dE = c_{VE} \tag{13}$$

where c_{VE} is the number of valence electrons per lattice site. As pointed out in Sect. B the AB^{IIIa} Zintl phases seem to be defect phases (see Table 4) and c_{VE} is not equal to the ideal electron-to-atom ratio. To calculate the Fermi energy for these non-perfect lattices the rigid band model is used. It is assumed that for all phases A_xB_{1-x} the shapes of the energy bands $E_n(\vec{k})$ are the same. Then the density of states curve is also the same for all alloys and based on this approach the alloys A_xB_{1-x} differ in the position of the Fermi energy. According to Eq. (13) E_F is a function of c_{VE}:

$$E_F = f(c_{VE}) \tag{14}$$

and c_{VE} is a function of x, (see Table 4):

$$c_{VE} = g(x) \; . \tag{15}$$

In Fig. 5 the integral over the density of states as a function of E is displayed. Following the rigid band model this integral is equal to c_{VE} (cf. Eq. (13)); if the defect structure is known, the range of the Fermi levels can be determined.

It is interesting to note that for $c_{VE} = 2.0$ the 5th and 6th bands of the AB^{IIIa} compounds are partly occupied. For the defect phases, however, c_{VE} is always smaller than 2 (cf. Sect. B). In this case the 5th and 6th bands are empty and the sheets of the Fermi surface are only formed by the 3rd and 4th bands[61, 63]. It seems that the occupation of the 5th and 6th bands is avoided by the formation of vacancies in the A sublattice and

Table 7. Fermi energies (in eV) and densities of states per eV and lattice site at the Fermi level for B32-type compounds AB deduced from the RAPW procedure[61–63]. (The values in brackets are gained from the ASA method[58, 59].) The bottom of the valence states is the zero of the energy scale. For the AB^{IIIa} compounds the Fermi energies E_F are listed for both the ideal stoichiometric compounds (c_{VE} is equal to the electron to atom ratio e/a = 2) and the compounds with defect structure, for which it is assumed that the fifths valence bands are non-occupied and E_F lies in the minimum of the density states curve, see Figs. 5–7. ($c_{VE} \approx 1.98$)

	c_{VE} = e/a			$c_{VE} \approx 1.98$		
AB	E_F	$N(E_F)$		E_F	$N(E_F)$	
LiZn	7.62	0.348	(0.312)			
LiCd	7.09	0.346	(0.318)			
LiAl	8.90	0.186	(0.260)	8.72	0.068	(0.101)
LiGa	10.53	0.124		10.44	0.040	
LiIn	9.41	0.192		9.26	0.063	
NaIn	8.29	0.198		8.06	0.134	
NaTl	8.92	0.145		8.75	0.134	

substitutionals in the B sublattice. Free energy considerations might suggest that the formation of these kinds of defects is more favoured than the filling of electronic states in the energy range of $N_5(E)$ which increases sharply (cf. Fig. 5).

In Table 7 the Fermi energies E_F and the densities of states at the Fermi level $N(E_F)$ are listed for all binary compounds covered. In order to study the defect structure of the AB^{IIIa} compounds, E_F and $N(E_F)$ are given for two valence electron concentrations $c_{VE} = 2.0$ and $c_{VE} \approx 1.98$. For the latter c_{VE} E_F lies in the minimum of the density of states curve and the 5th band is not occupied. Values of $N(E_F)$ obtained by the two different procedures (RAPW and ASA) are listed for LiAl, LiZn and LiCd. It seems that the differences in the $N(E_F)$ values are caused by differences in the hybridization effect in both methods (cf. Sect. C.II above).

Table 7 lists $N(E_F)$ for the AB^{IIb} and the AB^{IIIa} compounds, respectively. Whereas $N(E_F)$ is very small for the AB^{IIIa} compounds, the Fermi energies of the AB^{IIb} phases lie in the region of the maximum density of states which corresponds to $c_{VE} = 1.5$ in Fig. 5 (see also Fig. 6). It follows that the semimetal properties found for the AB^{IIIa} phases are less pronounced for the AB^{IIb} alloys.

C.IV. Charge Distribution

The chemical bonds in the B32 type Zintl phases are formed by the s and p valence electrons and by the outermost d electrons of the atoms of the groups IIb and IIIa in the Periodic Table. Based on the band structure calculations it was discovered[35, 59–61, 64] that the d-like bands are energetically separated from the valence electron bands. The values for the band gaps E_{ds} derived from the RAPW procedure are given in Table 8. According to the APW calculations for LiZn the valence and the d-like bands slightly overlap[62] whereas based on the ASA calculations[59] a band gap is also found for LiZn. It might be that the different choice of the exchange-correlation potential for both calculations is the origin of this discrepancy, as the position of the d-like bands depends strongly on the choice of the exchange parameter[35].

Table 8. Energy gaps E_{ds} between the d-like bands and the valence bands in B32-type compounds AB deduced from self-consistent RAPW calculations[61, 62] using the exchange-correlation potential of Vosko et al.[79]

AB	E_{ds}/eV
LiZn	0
LiCd	0.56
LiAl	–
LiGa	4.56
LiIn	4.98
NaIn	6.01
NaTl	1.46

To study the bonding mechanism it is helpful to consider the partial densities of state $q_y(E)$[82] which are the densities of states weighted by the amplitudes of the wave functions inside the atomic sphere y:

$$q_y(E) = N(E) \int_{\omega_y} \varrho(\vec{r}, E)d\tau \tag{16}$$

$\varrho(\vec{r}, E)$ is the charge density of the valence electrons associated to energy values E and ω_y is the volume of the atomic sphere being considered (cf. Table 5). The total charge of the valence electrons in the atomic sphere y is given by:

$$Q_y = \int_0^{E_F} q_y(E)dE . \tag{17}$$

Since the wave functions are expanded in spherical harmonics (Eq. (12)), $q_y(E)$ can be subdivided into l dependent components $q_y^{(l)}(E)$, $l = s, p, d, \ldots,$

$$q_y(E) = \sum_l q_y^{(l)}(E) . \tag{18}$$

The $q_y^{(l)}(E)$ are called l partial densities of states. Similarly, l partial charges $Q_y^{(l)}$ can be deduced from:

$$Q_y^{(l)} = \int_0^{E_F} q_y^{(l)}(E) . \tag{19}$$

The partial densities of states of LiAl and NaTl obtained from the RAPW calculations are shown in Figs. 8 and 9. For the lower two valence bands of the AB compounds it is found that $q_B(E)$ is much larger than $q_A(E)$: that is, the wave functions of these states possess much larger amplitudes in the B sublattice than in the A sublattice. Furthermore,

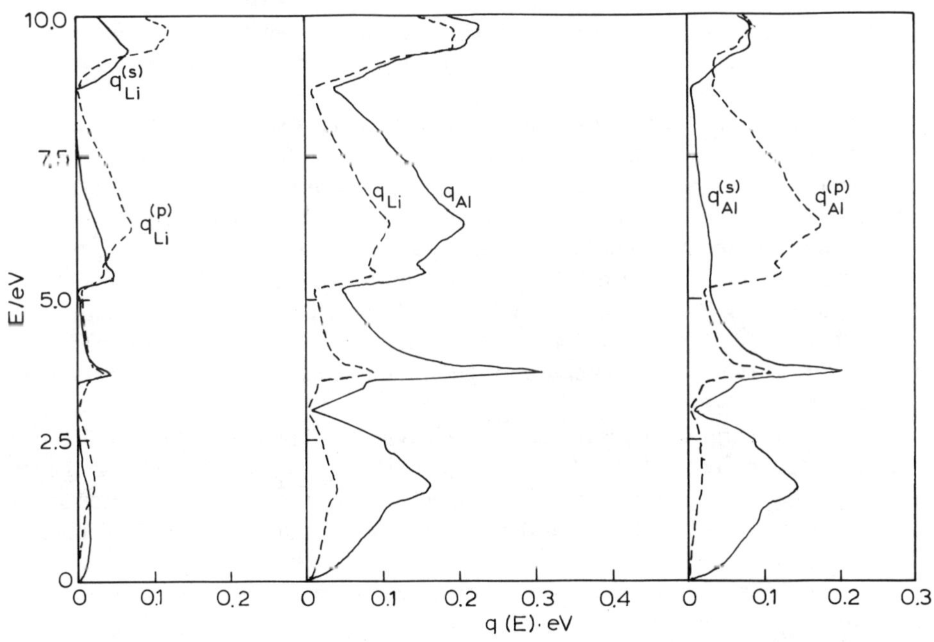

Fig. 8. Partial densities of states $q_y(E)$, y = Li and Al, for the first six valence bands for LiAl as a function of the energy E. The s and p partial densities of states $q_y^{(s)}$ and $q_y^{(p)}$ are shown too

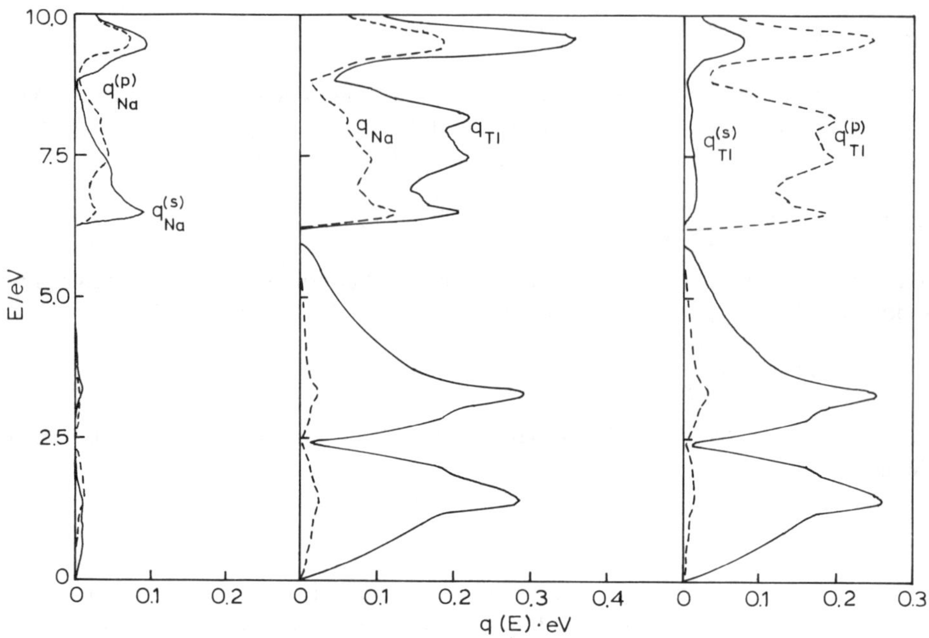

Fig. 9. Partial densities of states $q_y(E)$, y = Na and Tl, for the first six valence bands for NaTl as a function of the energy E. The s and p partial densities of states $q_y^{(s)}$ and $q_y^{(p)}$ are shown too

for these low lying valence bands $q_B^{(s)}(E)$ is the dominant l component. Therefore, the 1st and 2nd valence bands are predominantly formed by the atomic s valence electrons of the non-alkali atoms.

For the upper valence bands (3rd and 4th bands) q_A is still smaller than q_B but the differences between q_A and q_B are much smaller than for the lower valence bands. For the upper valence bands the p components of the wave functions have the largest amplitudes. One thus concludes that the upper valence bands are predominantly s-p-hybrid bands formed by both the alkali and the non-alkali atoms.

Figure 10 shows the d partial densities of states for LiAl and NaTl. One notices that the d components become important for upper valence states and for states of the conduction bands. The contributions of angular momenta larger than 2 are much smaller. More details about the l dependence of the different electronic states can be found in Zunger[57)] for LiAl.

The partial charge densities for all other B32 type Zintl compounds look similar to the ones displayed in Figs. 8–10. In Table 9 the partial charges Q_y and $Q_y^{(l)}$ calculated by the RAPW procedure[61–63)] are summarized for all binary compounds considered in this paper. In order to follow trends in the partial charges in the series of compounds the ratios $Q_y^{(l)}/Q_y$ are listed. We have not given the absolute values $Q_y^{(l)}$ because these values depend on the muffin-tin radii which are different for the various compounds (cf. Table 5).

Table 9 shows that the p-like charge dominates inside the Li atomic spheres. In the valence bond picture one can formulate a Li $s^{0.4}p^{0.6}$ valence electron configuration. For

Table 9. Charge analysis of the electronic valence states of intermetallic Zintl compounds AB from RAPW calculations[61, 62]. The total charges Q_y inside the atomic spheres y and the percentage l-partial charges $Q_y^{(s)}$ and $Q_y^{(p)}$ are listed

AB	y	Q_y/e^-	$Q_y^{(s)}/Q_y$	$Q_y^{(p)}/Q_y$
LiZn	Li	0.49	0.38	0.55
	Zn	1.38	0.59	0.38
LiCd	Li	0.64	0.38	0.55
	Cd	1.35	0.54	0.38
LiAl	Li	0.72	0.33	0.59
	Al	1.95	0.42	0.51
LiGa	Li	0.59	0.32	0.58
	Ga	1.97	0.51	0.47
LiIn	Li	0.66	0.34	0.56
	In	1.92	0.51	0.45
NaIn	Na	0.46	0.46	0.42
	In	2.04	0.55	0.42
NaTl	Na	0.47	0.48	0.39
	Tl	2.35	0.59	0.38

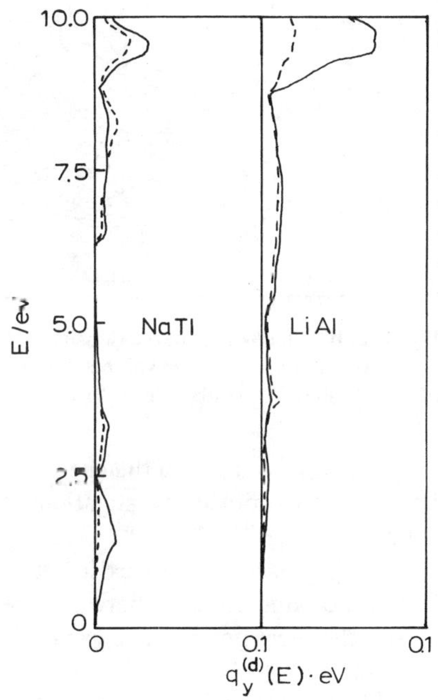

Fig. 10. d partial densities of states $q_y^{(d)}$ as a function of the energy E for Li *(dotted line)* and Al *(fully drawn line)* in LiAl and for Na *(dotted line)* and Tl *(fully drawn line)* in NaTl

Na one gets a much larger s character ($\approx 50\%$) than for Li. Moreover, for Na a d character of about 10% is found. One obtains almost the same l-like distribution for Zn and Cd as for Na. For the trivalent atoms the s-amplitudes increase with increasing atomic numbers. This is a general trend observed in molecular systems[83].

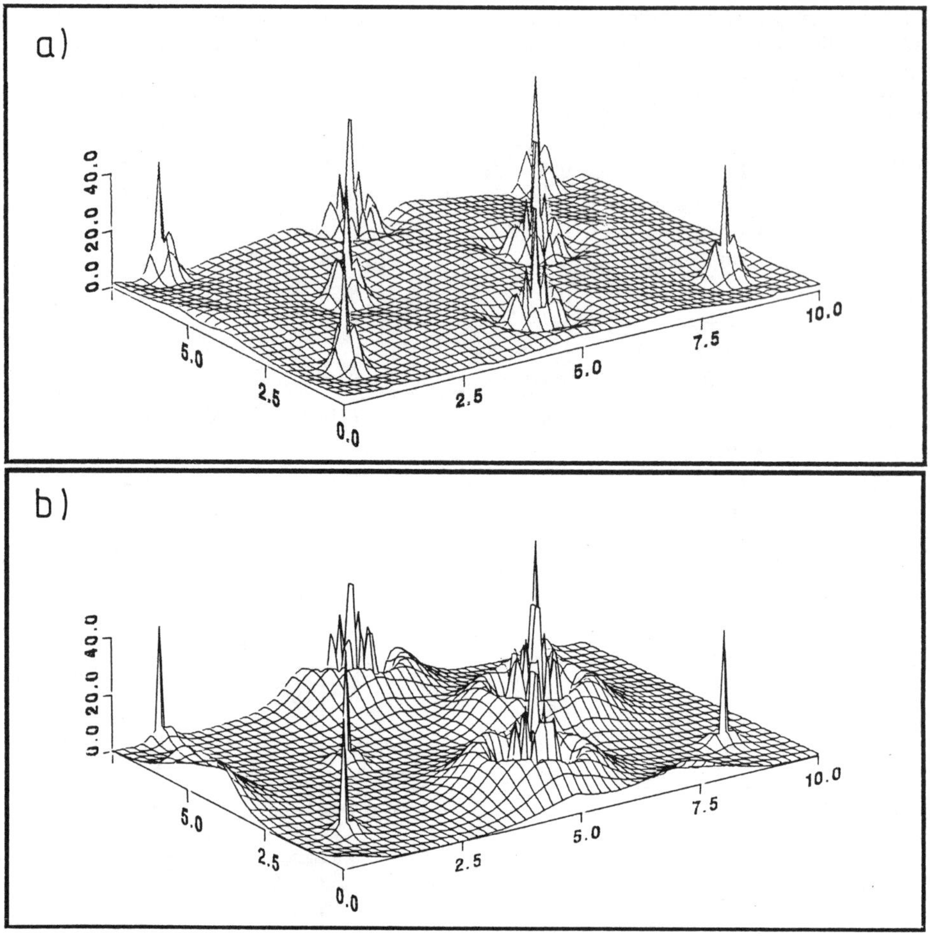

Fig. 11a, b. Electronic charge density $\varrho(\vec{r})$ in units of $-100\,e\,\text{Å}^{-3}$ for the (101) plane (see Fig. 4) for NaTl. **a)** $\varrho(\vec{r})$ for the upper valence bands (3rd and 4th bands in Fig. 5), **b)** $\varrho(\vec{r})$ for the lower valence bands (1st and 2nd bands in Fig. 5)

Finally, the nature of the chemical bond in the B32 type compounds is to be analyzed. Based on theoretical investigations the same bonding mechanism can be found for all compounds $AB^{IIb, IIIa}$ [57–63, 64]:

1. Strong covalent bonds exist between the atoms of the $B^{IIb, IIIa}$ sublattice;
2. non-bonding contacts between the alkali atoms are found;
3. the chemical bond between the two sublattices is covalent with a small ionic component; and
4. additionally, for the upper valence/conduction electron states a partial metal like charge distribution can be identified.

This bonding mechanism will be discussed in some detail in the remaining part of this subsection and in Sect. C.V.

Figure 11 displays the charge density of various electronic bands of NaTl in the (101) plane and Fig. 12 shows the total charge density. Figure 11d gives the charge density for

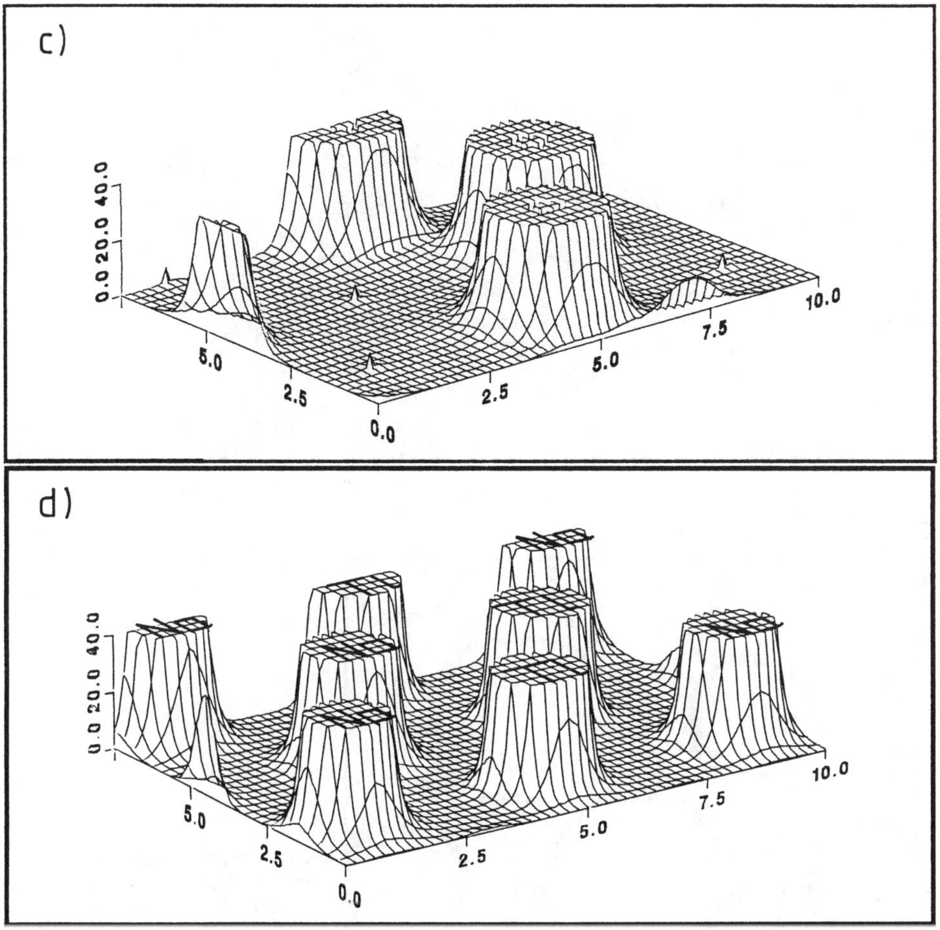

Fig. 11. c) $\varrho(\vec{r})$ for the 5d-bands of Tl, **d)** $\varrho(\vec{r})$ for the core electrons

the electronic core states (Na[1s... 2p]); Tl([1s...5p]). Figure 11c displays the overlapping 5d bands of Tl, which are more or less localized inside the Tl sublattice. In Fig. 11b the charge density of the lower two valence bands is plotted. From the shape of the charge density one recognizes strong covalent bonds in the Tl sublattice. The electron distribution of these bands is similar to the one in tetravalent semiconductors in the diamond-like structure. One notices a large gradient in the charge density going from the Tl to the Na sublattice for these lower states of NaTl. This gradient can also be seen from the partial density of states (see above).

In Fig. 11a $\varrho(\vec{r})$ is plotted for the two upper valence bands. Distinct differences in the electron distribution for these electronic bands compared to the ones of the lower valence bands (Fig. 11b) can be found, although a gradient in the charge density from the non-alkali to the alkali sublattice still exists. For the electronic states of the upper valence bands, however, the chemical bond possesses a metallic-like component. This metallic

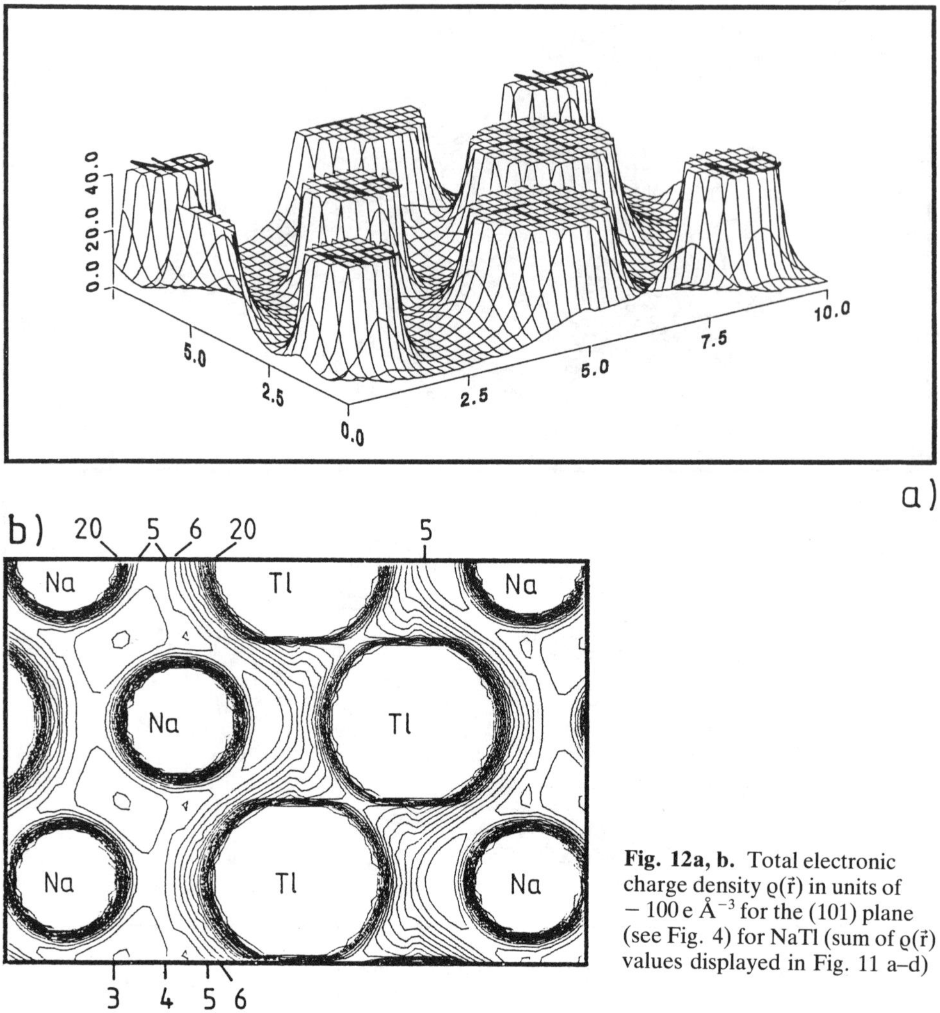

Fig. 12a, b. Total electronic charge density $\varrho(\vec{r})$ in units of $-100\,e\,Å^{-3}$ for the (101) plane (see Fig. 4) for NaTl (sum of $\varrho(\vec{r})$ values displayed in Fig. 11 a–d)

character is indicated in Fig. 11a by the fact that the bonding directions along the Tl–Tl nearest neighbors distances are less pronounced than in Fig. 11b and that the charge is more uniformly distributed over the whole crystal than in Fig. 11b.

C.V. Charge Transfer

As pointed out in the introduction a possible charge transfer in Zintl phases from the alkali to the non-alkali sublattice is often discussed in the literature. Therefore, there have been many attempts to calculate the charge transfer in these phases from band structure data. There is, however, no unique way to calculate a charge transfer. The various methods used and the resulting values for the Zintl phases are summarized in this subsection.

Theoretically, the charge transfer can be studied by comparing the charge density gained from the band structure ϱ_{band} with the charge density ϱ_{AO}, which is calculated by superposing the electronic densities of the free atoms without any charge relaxation. For the valence electrons of LiAl the difference $\varrho_{band} - \varrho_{AO}$ is plotted in Fig. 13 for the (110) plane.

From Fig. 13 one sees that in the region between the alkali atoms the redistribution of the charge in the crystal results in a charge reduction (ϱ_{AO} is larger than ϱ_{band}). Therefore, one finds non-bonding alkali-alkali contacts. The alkali- and non-alkali contacts are covalent/metallic (cf. above) with a slight charge deficit on the alkali side. A small ionic contribution to the bonding between the alkali and the non-alkali sublattices results from this deficit.

Besides a charge transfer from the alkali to the non-alkali sublattice one finds a delocalization of the electronic states of the d-like bands (Fig. 11c). A charge reduction ($\varrho_{band} - \varrho_{AO} < 0$) inside the non-alkali muffin-tin spheres is the result. This effect is not shown in Fig. 13.

Let us next turn to the quantitative results for the charge transfer. The charge transfer can be easily defined if localized LCAO wave functions are used. In this case the charge transfer can be determined on the basis of the occupation numbers of the atomic orbitals. For the Zintl phase LiAl Zunger[57] found $\Delta Q = 0.07\,e$ for the alkali sublattice. The problem in defining unambiguouisly the charge transfer by differences in the occupation numbers arises from the fact that this quantity depends on the chosen wave functions[57, 84].

On the other hand, the charge transfer can also be defined by the integral:

$$\Delta Q = e \int_{\Omega} (\varrho_{band} - \varrho_{AO}) d\tau \tag{20}$$

where Ω is the atomic volume. In this method, however, the choice of the atomic volume Ω is not well defined. For LiAl Asada et al.[58] have integrated over the Wigner-Seitz spheres and found a charge transfer of $\Delta Q = -0.02\,e$, which is again numerically small. This charge transfer, however, is opposite in sign to the values given above and contrary to chemists' experience for Li compounds[58].

Applying the integration over the muffin-tin spheres within the RAPW procedure the ratios Q_{band}/Q_{AO} have been determined (cf. Table 10). From these results one would predict a charge transfer of about 10% for the LiB compounds and more than 20% for the NaB compounds. As the total valence electron charge inside the alkali muffin-tin spheres is about $-0.6\,e$ (cf. Table 9) the APW result for the charge transfer in LiAl is close to the one obtained from the analysis of the occupation numbers (cf. above).

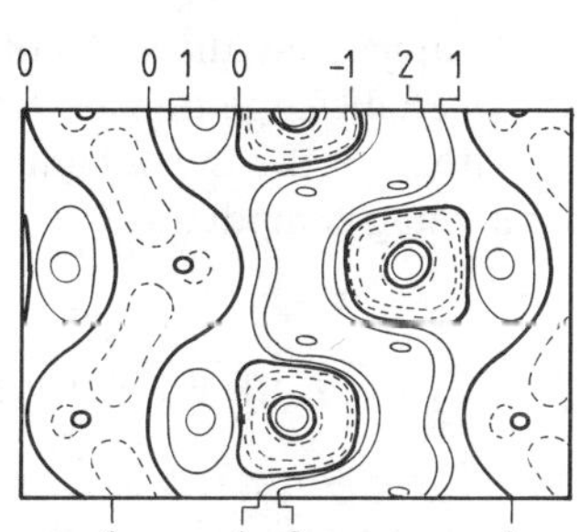

Fig. 13. Differences in the charge density $\Delta\varrho = \varrho_{band} - \varrho_{AO}$ (in units of $-100\,e\,Å^{-3}$ for LiAl for the (101) plane (see Fig. 4). ϱ_{band} and ϱ_{AO} are band structure results and the density gained from superposing the charge density of the free atoms, respectively

Table 10. Ratios of the electronic charge Q_{band}/Q_{AO} in B32 type compounds calculated from the spherical mean of the charge density inside the atomic volumes of the alkali atoms of radii $R = r_y$ (see Table 5) and $R = R_{NNN}$; $R_{NNN}(Li) = 1.00$ Å and $R_{NNN}(Na) = 1.35$ Å. Q_{band} is the charge determined from the band structure calculations and Q_{AO} is obtained from superposing the charge densities of the free atoms

	$R = r_y$	$R = R_{NNN}$
LiZn	0.91	0.83
LiCd	0.99	0.91
LiAl	0.91	0.78
LiGa	0.88	0.72
LiIn	0.89	0.73
NaIn	0.73	0.52
NaTl	0.79	0.55

Furthermore, a correlation between the charge transfer and the electronegativity of the B atom[62] is found for the Li containing compounds LiB.

The integration over spheres, however, can also only reflect trends in the charge transfer. As can be seen from Fig. 13 the gain in electron charge of the "anionic" sublattice is not concentrated in the muffin-tin spheres of the non-alkali atoms but in the whole sublattice.

Furthermore, there is the uncertainty in defining the boundary between the alkali and the non-alkali sublattice. From quantum mechanic arguments the atomic subsystems or atomic regions should be defined by surfaces given by[85]:

$$\vec{\nabla}\varrho\vec{n} = 0$$

where $\vec{\nabla}\varrho$ is the gradient of the electron charge density and $\vec{n}(\vec{r})$ is the normal of the surface. In this picture the alkali sublattice is much smaller than the non-alkali sublattice as can be seen from the position of the minima in the charge density shown in Fig. 12. The distances from an alkali nucleus to the minima in the charge density ϱ in the direction of the nearest non-alkali neighbor (R_{NN}) and next-nearest non-alkali neighbor (R_{NNN}) are $R_{NN}(Li) = 0.95$ Å and $R_{NNN} = 1.00$ Å for all Li containing compounds, and $R_{NN}(Na) = 1.26$ Å and $R_{NNN}(Na) = 1.35$ Å for NaIn and NaTl. Integrating the electron charge over spheres up to R_{NNN} one obtains the charge transfer given in the last column of Table 10 in percentage. These values are larger than those for $R = r_y$. However, the total valence electron charge inside the spheres with radii R_{NNN} are only -0.2 e for LiB and -0.25 e for NaB.

Summarizing the study of charge transfer in the B32 type compounds one finds that the overall charge transfer from the alkali to the non-alkali sublattice is less than -0.1 e per atom for the Li containing compounds and less than -0.2 e per atom for the sodium containing compounds.

C.VI. Ternary Systems

Electronic properties of ternary B32 type Zintl phases have been studied experimentally along the quasibinary intersection[17,18,35])

$AB_{1-x}C_x$, $0 \leq x \leq 1$.

In theoretical studies the rigid band model discussed above has been used over the entire range of composition for which the B32 structure type exists. As the rigid band model only works in special cases the quality of this model for the ternary Zintl phases investigated experimentally is to be discussed. For the systems $AB_{1-x}C_x$ the valence electron concentration c_{VE} as a function of x is given by:

$$c_{VE} = 0.5(n_A + (1 - x)n_B + xn_C)$$

where n_A, n_B and n_C are the numbers of the valence electrons of the atoms A, B and C respectively. The Fermi energy as a function of x is given by Eq. (13):

$$\int_0^{E_F(x)} N(E)dE = c_{VE}(x) \tag{21}$$

where N(E) is the density of states of one binary compound AB or AC. To study trends in the chemical properties as a function of composition within the rigid band model it is important that both phases AB and AC crystallize in the B32 structure and that the density of states for both phases in almost the same. In Fig. 14 the densities of states for the binary Li containing B32 type Zintl phases are plotted. In contrast to Fig. 6, however, N(E) is now a function of the electron concentration c_{VE}, as c_{VE} is the important quantity in the rigid band model.

One can see from Fig. 14 that in ternary systems like $LiCd_{1-x}In_x$ the rigid band model is a good approximation for those electronic properties which depend predominantly on

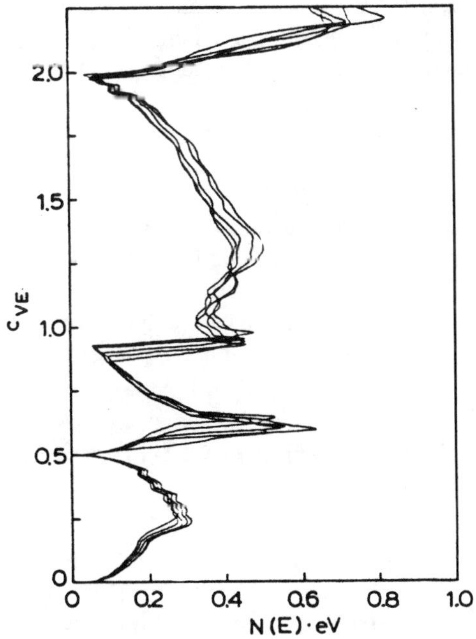

Fig. 14. Density of states N(E) for the LiB compounds as in Fig. 6. In this figure, however, N(E) is plotted as a function of the number of valence electrons per lattice site c_{VE} (as a function of the integral over N(E), $\int_0^E N(E')dE'$)

Table 11. Partial charges Q_y and $Q_y^{(l)}$ for Li_2CdIn. The first group of values (A) are gained from the band structure data of LiCd using the rigid-band model to get the number of valence electrons of Li_2CdIn. The second group of values (B) are those calculated similarly from the band structure data of LiIn. The third group of values (C) are the means of the values of A and B. The last group of values (D) are deduced from the band structure data of Li_2CdIn

	y	Q_y	$Q_y^{(s)}$	$Q_y^{(p)}$
A	Li	0.628	0.235	0.351
	Cd/In	1.331	0.716	0.504
B	Li	0.458	0.185	0.239
	In/Cd	1.480	0.933	0.510
C	Li	0.543	0.210	0.295
	Cd/In	1.405	0.824	0.507
D	Li	0.549	0.212	0.299
	Cd	1.324	0.716	0.441
	In	1.527	0.885	0.606

the density of states. It is found, however, that the model fails for those systems for which one parent metal B or C possesses electronic d-states which overlap strongly with the valence states[35)].

To study the influence of a third component on the charge distribution the band structure of the ordered Zintl phase Li_2CdIn has also been investigated. The structure of the ordered Zintl phase has been chosen even though Li_2CdIn crystallizes in the B32A type of structure (cf. Sect. B), because the band structure could only be determined for stoichiometric compounds and because only trends in the differences in the charge distribution of binary and ternary Zintl phases were investigated.

In Table 11 the results for the partial charges are listed. The muffin-tin radius is 1.438 Å for all calculations. The values given in Table 11 have been determined in two ways. Firstly, the partial charges for Li_2CdIn have been determined on the basis of the self-consistent band structures of LiCd and LiIn filling the electronic states up to the number of valence electrons of $c_{VE} = 1.75$, and using the lattice constant of Li_2CdIn ($a = 6.742$ Å[35)]). Secondly, the results from the band structure of ordered Li_2CdIn are presented. All values for the Li atomic spheres agree with each other quite well, whereas for Cd and In one finds that the charge inside the In sphere is larger in Li_2CdIn than expected according to the rigid band model. It is understandable that In has more charge than expected from the rigid band model as In is more electronegative than Cd (cf. Ref. 62). Obviously this interaction between the Cd and the In atoms is not taken into account in the rigid band model where the charge distribution of Li_2CdIn is only deduced from band structure data of the binary phases LiCd and LiIn.

D. Comparison of Electronic States in the B2 and the B32 Structure

The number of AB compounds which crystallize in the B2 type of structure is much larger than those which form B32 type Zintl phases. As pointed out in Section B the NaTl type is restricted to alloys for which the sizes of the parent atoms are almost equal and for which the valence electron concentration is restricted to $1.5 \leq c_{VE} \leq 2.0$. To gain some insight into the differences between the bonding mechanisms in the structures B2 and B32, the band structures of a few compounds were calculated for both arrangements[35, 61, 62, 64, 65]. In these calculations the band structure of the hypothetical phase is always performed for lattice constants:

$$a_{B32} = 2\, a_{B2} \; . \tag{22}$$

Under this condition the volumes of the crystals are the same in both theoretical investigations and the lattices differ only in the arrangement of the atoms on the lattice sites of the bcc lattice (cf. Fig. 1).

In Sect. D.I the differences in the electronic states of both structures are discussed and in Sect. D.II differences in the total energies:

$$\Delta E = E_{B32} - E_{B2} \tag{23}$$

are considered.

D.I. Comparison of the Band Structures

The tendency of diminution of physical properties from "metallic" to "semi-metallic" properties observed for intermetallic Zintl phases (cf. Sec. A) could not be found in B2 type compounds. The differences in the symmetry between the two structures seem to be important for this behavior. In Fig. 15 the density of states N(E) of LiCd is shown. For comparison N(E) is also given for a hypothetical B2 type phase LiCd(B2). N(E, LiCd(B2)) shows fewer peaks than N(E, LiCd) and the low minimum of N(E) in the range of the Fermi energy of the AB^{IIIa} compounds (cf. Figs. 5 to 7) is missing for the B2 type compounds[61, 62, 86]. For the upper valence states the shape of N(E, B2) is more free electron like than N(E, B32) and this is caused by a large overlap of the different bands in this energy range for the B2 type of structure[61].

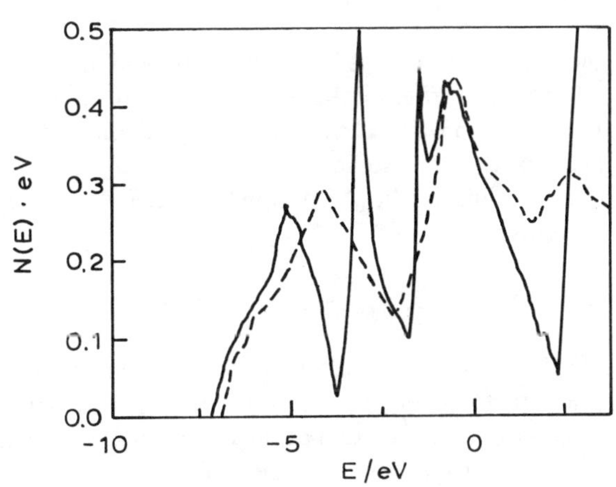

Fig. 15. Density of states for the first six valence bands for LiCd *(fully drawn line)* and for the first three valence bands of hypothetical LiCd(B2) *(dashed line)* as a function of E. The zero of the energy scale is the Fermi energy (Schmidt[62])

Table 12. Charges inside the atomic spheres Q_y and the ratios $Q_y^{(l)}/Q_y$ of the l partial charges for LiTl and NaTl. For both compounds the self consistent scalar relativistic APW calculations have been performed for both structures B2 and B32. The lattice constants are chosen $a_{B32} = 2\ a_{B2}$ where $a_{B32}(NaTl) = 7.47$ Å and $a_{B2}(LiTl) = 3.42$ Å are the lattice constants found experimentally. The volumes of the muffin-tin sphere are $\omega_{Li}(LiTl) = \omega_{Tl}(LiTl) = 12.47$ Å^3 and $\omega_{Na}(NaTl) = \omega_{Tl}(NaTl) = 16.84$ Å^3

AB	y	Q_y	$Q_y^{(s)}/Q_y$	$Q_y^{(p)}/Q_y$
LiTl(B2)	Li	0.712	0.33	0.57
	Tl	1.891	0.62	0.33
LiTl(B32)	Li	0.594	0.35	0.56
	Tl	2.024	0.56	0.37
NaTl(B2)	Na	0.597	0.50	0.38
	Tl	2.204	0.65	0.33
NaTl(B32)	Na	0.433	0.47	0.41
	Tl	2.290	0.60	0.37

The calculated charge distribution depends also remarkably on the type of structure chosen. In Table 12 the partial charges Q_y, Eq. (17), for LiTl and NaTl are listed for the B2 and the B32 structures. Considering the alkali partial charges one finds that $Q_A(B2)$ is always larger than $Q_A(B32)$. The Tl spheres exhibit the opposite behaviour. This indicates that the charge is more uniformly distributed in the B2 type of structure than in the case of B32. The larger metallic component in the chemical bonds for the B2 phases can also be obtained from contour plots[61)]. Comparing the s and p character of the electronic states for both types, no distinct differences for the alkali atoms can be found whereas for Tl the s character is larger in the B2 type phases than in those of B32.

The charge transfer ΔQ has also been studied as a function of the structure chosen. It was discovered[61)] that ΔQ depends on the structure in such a way that ΔQ is smaller for the B2 type than for the B32. For LiTl(B2) only a small charge transfer is calculated. Our findings also agree with experimental investigations of intermetallic molecules[87)] where a larger charge transfer is found for the NaTl molecule than for LiTl.

D.II. Energy Differences

The energy differences between the B2 and the B32 structures Eq. (23) can be calculated either by perturbation theory[54–56)] or by calculating E(B2) and E(B32) separately[61, 62, 64)]. In this section we shall first briefly outline the different methods to calculate ΔE and then give the results of some theoretical investigations.

According to the APW approach ΔE (Eq. (23)) can be written as the sum of four elements:

$$\Delta E = \Delta E_{band} + \Delta E_{local} + \Delta E_{xc} + \Delta E_{out} \ . \qquad (24)$$

The first term is the difference in the sum of the one-electron energies deduced from band structure calculations for the B32 and B2 structures, respectively. The second term is a coulombic interaction of the charge densities inside the muffin-tin spheres, ΔE_{xc} is an exchange-correlation correction, and ΔE_{out} is an interatomic energy difference which includes differences in Madelung terms. As mentioned above the calculations were performed under the condition that $a_{B32} = 2\ a_{B2}$ (Eq. (22)). From arguments given in Refs. 90 and 91 it follows that in this case ΔE can be derived more efficiently from the first term in Eq. (24) by carrying out calculations for the B2 and B32 phases with the same muffin-tin potential. If $N_{B32}(E)$ and $N_{B2}(E)$ are the densities of states for the B32 and the B2 phases, one obtains:

$$\Delta E'_{band} = \int_0^{E_F^{B32}} N_{B32}(E)dE - \int_0^{E_F^{B2}} N_{B2}(E)dE\ . \tag{25}$$

The prime at $\Delta E'_{band}$ indicates that in this approximation the self-consistent potential is calculated only once and that the band structures $E(\vec{k}, B2)$ and $E(\vec{k}, B32)$ are thus different due to the differences in the structure factors. In this model ΔE is determined in a procedure similar to the Hückel calculations for organic molecules.

In the atomic sphere model (ASA) one gets to a good approximation[64, 92]:

$$\Delta E = \Delta E'_{band} + \Delta E_{as} + \Delta E_{mad} \tag{26}$$

where ΔE_{as} is an intraatomic Coulomb term and ΔE_{mad} is the difference in Madelung energies. In a semiquantitative discussion ΔE_{as} can be omitted.

It should be pointed out that the energy expressions in Eqs. (25) and (26) – especially the Madelung terms – are not the same, because the atomic spheres overlap in the ASA model whereas in the APW approach the muffin-tin spheres only touch.

$\Delta E'_{band}$ can also be calculated from perturbation theory via the pseudo-potential matrix elements[54–56]. The pseudo-potential approach, however, is only justified if the parent metals have the same valency and the same Fermi vector, and if furthermore no charge transfer takes place and the AB compounds have the same atomic volumes as A and B[56].

After having discussed the three approaches, let us now turn to the results of the calculations. The values for $\Delta E'_{band}$ were calculated for all binary B32 type Zintl compounds and for the B2 type phase LiTl[57, 61, 62, 64]. In the perturbation approach[57] and the ASA method[64] $\Delta E'_{band}$ always favours the B32 type of structure (also in case of the B2

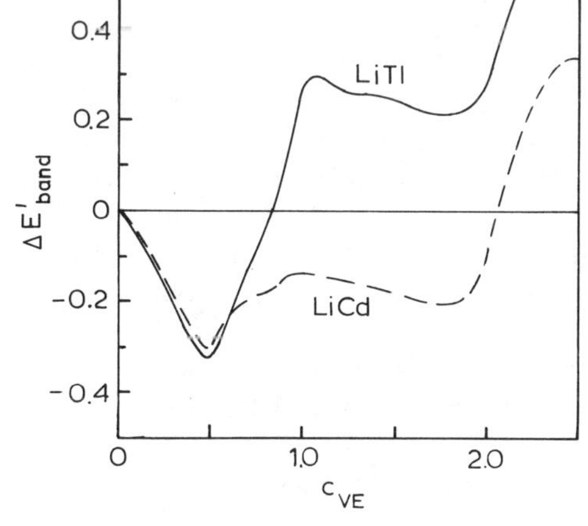

Fig. 16. Differences in the ordering energy $\Delta E'_{band}$ between the B2 and the B32 type of structure as a function of the number of occupied states (number of valence electrons per lattice site c_{VE}) for LiCd and LiTl calculated from the one-electron energies by use of the self-consistent muffin-tin potential of LiTl and LiCd respectively

type compound LiTl). In the RAPW procedure $\Delta E'_{band}$ favours the structure which is experimentally found[61, 62].

In order to study the influence of electronic effects on the stability of binary and ternary Zintl phases within the rigid band model (Sect. C.VI) Fig. 16 shows $\Delta E'_{band}$ of LiTl and LiCd as a function of the valence electron concentration. The maxima of the density of states curves for the B2 and the B32 types of structures occur at different energy values (cf. Fig. 15), and therefore $\Delta E'_{band}$ as a function of c_{VE} is an oscillating curve. The reasons why the curve for LiTl changes its sign are discussed below.

The shape of the curves in Fig. 16 suggests that the electronic states in the B2 and the B32 structures play an important role for the B2-B32 phase transition in the system $LiCd_{1-x}Tl_x$ (cf. Fig. 3). To study the influence of the electronic structure on the position of the phase transition as a function of x quantitatively, however, other mechanisms, such as self-consistency effects and the influence of the non-stoichiometry, have to be known.

To get some idea about possible causes for a B2-B32 transition due to electronic effects, the interatomic terms (ΔE_{out}, Eq. (24) and ΔE_{mad}, Eq. (26)) and the kinetic energy are considered.

The interatomic terms always favour the B2 type of structure, in accordance with the general finding that for ionic crystals all nearest neighbours are unlike ions. With decreasing lattice constant the Madelung energy decreases, and for B32 type compounds a B32-B2 phase transition might occur at high pressure conditions[64]. This pure geometric effect is modulated by the charge redistribution. In the muffin-tin approximation used in the RAPW procedure the charge is more uniformly distributed in the B2 phase than in the case of B32 (cf. Table 12 above), and the geometric effect is reduced by the charge redistribution. In the ASA picture the Madelung term in Eq. (26) is enhanced because in this model the "charge transfer" is larger for the B2 than for the B32 phase[64].

In addition to the influence of the interatomic terms the role of the internal pressure P shall be considered. The virial theorem can be written as[88, 89]:

$$\frac{3}{2}\,P\omega = T + \frac{1}{2}\,V' \tag{27}$$

where V' is the potential energy V corrected by an exchange-correlation term. P increases with decreasing interatomic distances, and as a function of interatomic distances the kinetic energy T can increase more rapidly than -2 V. Because of the condition that $a_{B32} = 2\,a_{B2}$ the interatomic distances of the same atoms are smaller in the B32 type of structure than for B2. It follows for LiTl that, as far as the space filling arrangement in the B32 structure is concerned, the Tl atoms are so compressed that the internal pressure results in $\Delta E > 0$ and also $\Delta E'_{band} > 0$[61]. This seems to be the reason why the atomic sizes of all atoms present in B32 type compounds are roughly the same. In NaTl the alkali atoms are larger. In this case, however, no drastic differences are found for the kinetic energy in the B2 and the B32 structures[61]. This is caused by the charge transfer from the Na to the Tl sublattice (cf. Table 10). This charge transfer is significantly larger for the Na compounds than for the Li systems. This can not be explained by differences in the electronegativities. It might be that there is a correlation between the charge transfer and the compression of the atoms.

E. Optical and Magnetic Properties

The optical and magnetic properties of some intermetallic Zintl phases differ from those which are commonly found for intermetallic phases. The ternary systems $LiAg_{1-x}In_x$, $LiCd_{1-x}In_x$ and $LiZn_{1-x}Cd_x$ are colored, and the color changes as a function of x continuously from yellow to blue[35]. This change in color is due to shifts of the minima in the reflectivity as a function of x for photon energies $E = \hbar\omega$ within the optical range[35]. In a microscopic theory, band structure data are not directly correlated with the reflectivity. In a simple approximation, however, the imaginary component of the dielectric function $\varepsilon_2(\omega)$ can be calculated from the dispersion curves $E_n(\vec{k})$[93]. As an example, the experimental and the theoretical results for the optical constants in the system $LiCd_{1-x}In_x$ are reported in Sect. E.I.

The following magnetic properties have been measured in B32 type Zintl phases: the magnetic susceptibility χ and the Knight shift K_s. Internal magnetic fields must be traced back to the magnetic spin moments and the movement of the electrons (orbital contributions). The theoretical investigation of χ and K_s is rather complicated[94–98] and simplifying assumptions have to be used in order to investigate χ and K_s theoretically. In these models χ and K_s are separated into various terms. In Sects. E.II and E.III these terms will be discussed. It will be demonstrated that most experimental data can be well understood on the basis of the electronic properties of the Zintl phases discussed in Sect. C.

E.I. Optical Properties

Colored Zintl phases, which crystallize in the NaTl type of structure, are found for the alloys $Li_2Cd_{2-x}In_x$, $0.6 \le x \le 1.0$ and $Li_2Zn_{2-x}Cd_x$, $1.2 \le x \le 1.6$. The optical constants (reflectivity r, "polarization" ε_1 and "absorption" ε_2) were determined[35] for these sys-

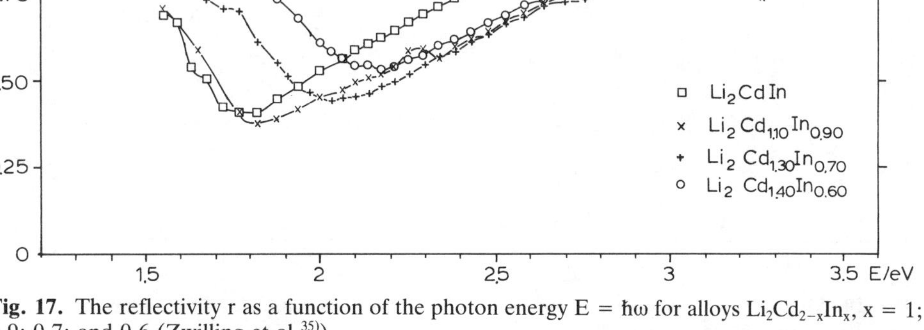

Fig. 17. The reflectivity r as a function of the photon energy $E = \hbar\omega$ for alloys $Li_2Cd_{2-x}In_x$, x = 1; 0.9; 0.7; and 0.6 (Zwilling et al.[35])

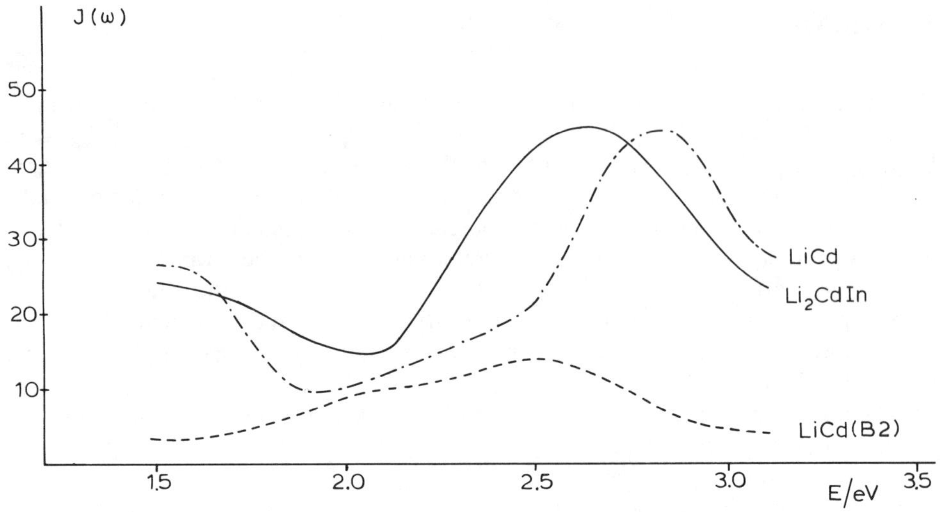

Fig. 18. Joint density J(ω) as a function of E = ħω for LiCd and Li_2CdIn. For Li_2CdIn the results for the band structure of LiIn and the rigid band model were used. The *dotted line*, LiCd(B2), show J(ω) assuming the CsCl type of structure for LiCd. J(ω) is given in arbitrary units (Zwilling et al.[35])

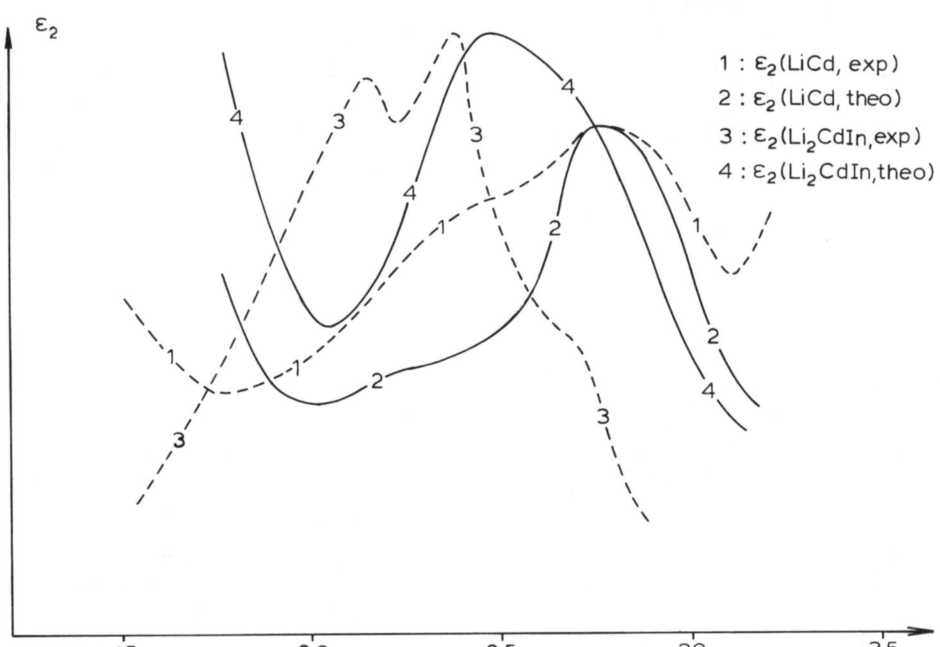

Fig. 19. Measured and calculated absorption coefficient ε_2 as a function of E = ħω for LiCd and Li_2CdIn. ε_2 is given in arbitrary units (Zwilling et al.[35])

tems. Figure 17 shows the measured reflectivity r as a function of the photon energy $E = \hbar\omega$ for the system $Li_2Cd_{2-x}In_x$. r shows minima in the optical energy range and the minima in r shifts with decreasing x from 2.18 eV for x = 1.4 to 1.77 eV for x = 1.0.

Theoretical studies were performed for ε_2[35]. Using a simple microscopic theory ε_2 is correlated to direct interband transitions. $\varepsilon_2(\omega)$ can be calculated using a simple approach[35, 93] from the joint density of states $J(\omega)$, see below. $\varepsilon_2(\omega)$ is then:

$$\varepsilon_2(\omega) = C\,J(\omega)/\omega^2 \qquad (28)$$

where it is assumed that the unknown factor C does not depend on ω. $J(\omega)$ is defined by:

$$J(\omega) = \sum_{\vec{k},n,n'} \delta[E_{n'}(\vec{k}) - E_n(\vec{k}) - \hbar\omega] \qquad (29)$$

where $E_n(\vec{k})$ is the initial state below the Fermi energy E_F and $E_{n'}(\vec{k})$ is the final state above E_F. Figure 18 shows the joint density for LiCd and Li_2CdIn and for the hypothetical phase LiCd(B2). The hypothetical B2 phase LiCd(B2) has been included in these investigations to study the role of the structure on optical transitions. From Fig. 18 one sees that $J(\omega)$ shows sharp maxima in the energy range between 2.5 and 3.0 eV for LiCd and Li_2CdIn. For LiCd(B2) only a broad and flat maximum was obtained.

Figure 19 shows the experimental and theoretical results for ε_2 for LiCd and Li_2CdIn. ε_2(LiCd, exp) shows a maximum for $\hbar\omega = 2.8$ eV. This maximum shifts to lower photon energies for LiCdIn. Using the joint density one obtains a maximum for ε_2(LiCd, theor) which corresponds to the experimental value. This maximum in the theoretical curve is shifted to lower energy values for Li_2CdIn. This shift, however, is smaller than the one found experimentally.

ε_2 derived from $J(\omega$, LiCd(B2)) is not plotted in Fig. 19. As $J(\omega$, LiCd(B2)) possesses only a flat maximum in the optical energy range (Fig. 18) no maximum occurs for $\varepsilon_2(\omega, \text{LiCd(B2)}) = CJ(\omega, \text{LiCd(B2)})/\omega^2$. Based on the density of states N(E) given in Fig. 15 one can notice that the peaks in N(E) are less pronounced in the B2 than in the B32 structure. These differences in N(E) between the two structures are similar for the joint density.

The results of the optical investigations can be summarized as follows.

1. For colored Zintl phases a minimum in the reflectivity r_{min} and a maximum in the absorption $\varepsilon_{2,\,max}$ were obtained in the optical energy range $1.5 \leq \hbar\omega/\text{eV} \leq 3.5$.
2. For a series of compounds $A_2B_{2-x}C_x$ r_{min} and $\varepsilon_{2,\,max}$ shift to lower photon energies with increasing valence electron concentration of the alloys.
3. Theoretical values for $\varepsilon_{2,\,max}$ calculated for the system $Li_2Cd_{2-x}In_x$ in a simple way from the joint density show the same trend as the experimental finding. Furthermore, the theoretical studies suggest that the optical constants of these phases depend on the specific shape of the electronic bands in the NaTl type of structure.

E.II. Magnetic Susceptibility

The (dimensionless) volume susceptibility in SI units is given by:

$$\chi' = \mu_0 \frac{M}{B} \qquad (30)$$

where M is the magnetization per volume of the formula unit AB, μ_0 is the permeability and B is the magnetic induction. To obtain values for the susceptibility in the widely used cgs-units, one has to include a factor of $1/4\pi$:

$$\chi = \chi'/(4\pi) \tag{31}$$

The molar susceptibility χ_{mol} can be obtained from:

$$\chi_{mol} = N_A \omega_{AB} \chi$$

where ω_{AB} is the volume per formula unit AB for binary Zintl phases and N_A is Avogadro's number.

In the first three columns of Table 13 the experimental values for the susceptibility χ_{mol} are listed for the binary B32 type Zintl phases[25, 26]. Column 4 gives the contributions of the core electrons based on the density functional approach. In the last two columns of Table 13 the contributions of the valence electrons are shown.

The following experimental findings are important: a) χ depends strongly on the alkali concentration, b) for LiCd and NaTl χ is dependent on the temperature, c) χ for NaIn is not well known, and furthermore there are differences between the results of different authors.

Theoretically, the total susceptibility can be separated into an orbital $\chi^{(l)}$ and a spin contribution $\chi^{(\S)}$. Firstly, $\chi^{(\S)}$ will be considered. Theoretical values for $\chi^{(\S)}$ are obtained from band structure calculations[58, 59, 61, 63]. These values correspond to a temperature of 0 K. $\chi^{(\S)}$ is due to the difference between the number of spin-up and spin-down electrons per formula unit AB, N_+ and N_-:

$$\chi^{(\S)} = \frac{\mu_0 \mu_B (N_+ - N_-)}{4\pi\, \omega_{AB} B} \tag{32}$$

Table 13. Experimental values for the molar susceptibility in binary B32-type compounds in units of 10^{-6} cm^3/mol. The first two columns are the values of Klemm and Fricke[25]. These are average values of 48 to 52 at. pct. of Li. The third column contains the values of Yao[26]. These are interpolated values for the composition of 50 at. pct. Li. The 4th column contains the contribution of the core electrons gained for density functional calculations for free atoms. In the 5th and 6th columns the contributions of the valence electrons are listed: $\chi_1(\text{val}) = \chi_{exp,1} - \chi_{core}$ and $\chi_3(\text{val}) = \chi_{exp,3} - \chi_{core}$. (Numbers in parentheses indicate error margins.)

	1	2	3	4	5	6
	$\chi_{exp,1}$ (90 K)	$\chi_{exp,2}$ (293 K)	$\chi_{exp,3}$ (293 K)	χ_{core}	χ_1(val)	χ_3(val)
LiZn	4 (1)	3	15	−13	17	28
LiCd	−11 (1)	−2	7	−25	13	32
LiAl	7	7	2	−4	11	6
LiGa			−13	−11		−2
LiIn	2 (1)	2	−29	−22	24	−7
NaIn	−28 (14)	−28		−27	−1 (14)	
NaTl	−116 (2)	−130 (2)		−36	−80	

Table 14. Spin susceptibility $\chi^{(s)}$ for B32-type compounds AB[61, 63]. For the AB^{IIIa} compounds the results are listed for the stoichiometric (st) and for the defect (de) phases. $\chi^{(s)}_{Pauli}$ is the Pauli susceptibility deduced from the density of states; $\chi^{(s)}$ is the spin susceptibility including the exchange enhancement; and $\chi^{(s)}_{mol}$ is the molar spin susceptibility in units of cm^3/mol

		$\chi^{(s)}_{Pauli} \cdot 10^6$	$\chi^{(s)} \cdot 10^6$	$\chi^{(s)}_{mol} \cdot 10^6$
LiZn		1.25	1.60	29
LiCd		1.00	1.26	28
LiAl	st	0.61	0.68	13
	de	0.25	0.28	5
LiGa	st	0.45	0.48	9
	de	0.25	0.24	4
LiIn	st	0.53	0.59	14
	de	0.19	0.21	5
NaIn	st	0.45	0.48	14
	de	0.22	0.24	7
NaTl	st	0.29	0.33	10
	de	0.27	0.31	10

where μ_B is the Bohr magneton. In the Pauli model[99] $N_+ - N_-$ is given by the density of states (per formula unit AB) at the Fermi level $N(E_F)$:

$$\chi^{(s)}_{Pauli} = \frac{\mu_0 \mu_B^2 N(E_F)}{4\pi\omega_{AB}} . \tag{33}$$

In this model, however, important exchange correlation effects are not considered[100]. To include such effects the author has performed spin polarized band structure calculations[78, 101] for B32 type compounds[61, 63]. Using this approach the one-electron equations are different for the spin-up and spin-down electrons[78, 101]. The magnetization $\mu_B \cdot (N_+ - N_-)$ can be calculated from the one-electron energies by summing up the number of occupied spin-up and spin-down electronic states. The resulting value for $\chi^{(s)}$ (Eq. (32)) is always larger than the Pauli susceptibility (Eq. (33)), because the exchange correlation enhances the number of singly occupied states at the Fermi level (exchange enhancement effect).

In Table 14 the data for the spin susceptibilities obtained from Eqs. (32) and (33) are listed. For the AB^{IIIa} compounds the data are given for the stoichiometric (st, $c_{VE} = 2$) and for the defect (de, $c_{VE} \sim 1.98$) phases (cf. Sects. B and C). As expected from the values of the densities of states at the Fermi surface (Table 7), the spin susceptibilities are small for the defect phases and the values $\chi(AB^{IIb})$ are about a factor of four larger than $\chi(AB^{IIIa})$. The exchange enhancement factors $\chi^{(s)}/\chi^{(s)}_{Pauli}$ are in the range of 1.1 for the AB^{IIIa} phases and roughly 1.3 for AB^{IIb}. The values in Table 14 are calculated from RAPW band structure data[61, 63]. For LiAl, LiZn, and LiCd the Pauli susceptibility is also determined from the ASA model[58, 59]. These values differ from the RAPW results due to differences in the density of states given in Table 7.

The difference between χ(val) (Table 13) and $\chi^{(s)}_{mol}$ (Table 14) should yield some information about the orbital contribution $\chi^{(l)} \approx \chi(\text{val}) - \chi^{(s)}_{mol}$. A comparison between χ(val) and $\chi^{(s)}_{mol}$ shows that for the AB^{IIIa} compounds $\chi^{(s)}$ and $\chi^{(l)}$ should be of the same order of magnitude. For the AB^{IIb} compounds it might be that $\chi^{(s)}$ dominates. An accurate calculation of $\chi^{(l)}$ has yet not been performed for the B32 type compounds.

E.III. Knight Shift

In NMR measurements the Knight shift K_s is defined as the shift of the resonance field with respect to the resonance field of the reference sample (ionic solution):

$$K_s = \frac{B_{ref} - B}{B_{ref}} \; . \tag{34}$$

Theoretically K_s is given by

$$K_s = \frac{\sum_i E_{hfs,\,i}}{\mu B} \tag{35}$$

where $\mu \cdot B$ is the nuclear Zeeman energy considered and $E_{hfs,\,i}$ is the expectation value of the hyperfine energy for the ith electron. From the hyperfine spin Hamiltonian one can derive two contributions to K_s, $K_s^{(\S)}$ and $K_s^{(\hat{l})}$, analogously to the susceptibility discussed in the previous section. The hyperfine field due to the spin moments of the electrons will be considered first. According to the non-relativistic approach one obtains a Fermi contact term[102] due to singly occupied states at the Fermi surface. This contribution is determined from:[58, 59, 61, 63]

$$K_s^{dir} = \frac{8\,\pi}{3}\, \chi_{Pauli}^{(\S)} < \langle |\psi_{\vec{k}_F}(\vec{R}_y)|^2 \rangle_F \omega \tag{36}$$

where $\vec{k}_F$ is a wave vector at the Fermi surface and $\vec{R}_y$ is the position of the nucleus considered. This direct contribution (Eq. (36)) has to be corrected by two exchange correlation terms. Firstly, one has to include the exchange enhancement of the susceptibility (cf. above). Additionally, the doubly occupied electronic states below the Fermi surface produce a hyperfine field at the nuclei due to the exchange polarization effect[61, 103, 104]. For theoretically investigating the Knight shift in B32 type Zintl phases by the RAPW method, the relativistic expressions for the hyperfine fields[86] were used[61, 63]. In Table 15 the spin contributions to K_s, the experimental results for K_s, and for comparison K_s for the pure metals, are listed. Only the s partial waves contribute to K_s in the non-relativistic treatment. Therefore, the ratios of the s partial density of states at the Fermi level E_F to the total partial density are given as well. Two values for the Knight shift are listed in Table 15. K_s^{dir} is the Fermi contact term, Eq. (36), and $K_s^{(\S)}$ includes exchange polarization and exchange enhancement effects.

As defect structures are found for the AB^{IIIa} Zintl compounds, the theoretical values $K_s(AB^{IIIa})$ are listed for both the stoichiometric (st) and the defect (de) phases (cf. Sects. B and C).

The experimental results will be considered first of all. Comparing the values for K_s found in the Zintl phases with the ones in the pure metals one notices that the AB^{IIb} phases show a "normal" paramagnetic behaviour. For the AB^{IIIa} compounds K_s is distinctly smaller than in the case of pure metals. K_s is even negative for NaTl.

Some insight into these differences in K_s^{exp} for the AB^{IIb} and AB^{IIIa} phases can be obtained from the calculated spin contributions to $K_s^{(\S)}$. The Fermi contact term is always

Table 15. Ratio of the s partial charge densities $q_y^{(s)}(E_F)$ to the total partial densities $q_y(E_F)$ for electronic states at the Fermi level E_F and Knight shifts K_s (in %) for intermetallic Zintl compounds AB[61, 63]. The calculated direct contribution $K_s^{dir.}$ and the total spin contribution $K_s^{(s)}$ are listed. For comparison the Knight shift found experimentally in the Zintl (K_s^{exp}) phases and in the pure metals (K_s^{met}) are given too. For the AB^{IIIa} compounds the theoretical values are shown for both the defect (de) and the stoichiometric (st) phases

		$q_y^{(s)}(E_F)/q_y(E_F)$		$K_s^{dir.}$		$K_s^{(s)}$		K_s^{exp} Ref. 32	K_s^{met} Ref. 32
AB	y	st	de	st	de	st	de		
LiZn	Li	0.20		0.010		0.013		0.00(1)	0.026
	Zn	0.13		0.18		0.15		0.20(5)	0.20
LiCd	Li	0.19		0.010		0.012		0.01	0.026
	Cd	0.14		0.32		0.25		0.39(2)	0.415
LiAl	Li	0.64	0.04	0.018	0.0003	0.020	0.001	0.005	0.026
	Al	0.31	0.09	0.09	0.02	0.09	0.01	0.01(3)	0.164
LiGa	Li	0.64	0.05	0.012	0.0005	0.015	0.001	0.005	0.026
	Ga	0.44	0.11	0.28	0.05	0.25	0.04	0.09(1)	0.146
LiIn	Li	0.65	0.01	0.018	0.0004	0.022	0.001	0.005	0.026
	In	0.42	0.12	0.63	0.09	0.58	0.06	0.13(2)	0.82
NaIn	Na	0.60	0.12	0.038	0.003	0.050	0.003		0.113
	In	0.20	0.07	0.33	0.10	0.30	0.08	0.07; 0.22	0.82
NaTl	Na	0.73	0.21	0.09		0.10	0.0025	−0.016	0.113
	Tl	0.23	0.10	1.31		0.99	−0.14	−0.96	1.67

positive and K_s^{dir} is the main spin contribution for the AB^{IIb} compounds. $K_s(A\underline{B}^{IIb})$ is large because the Fermi level for these compounds lies in the range of a large density of states and a large s partial density of states (cf. Sect. C.IV).

For the AB^{IIIa} compounds the values $K_s^{(s)}$ are quite different for the stoichiometric (st, $c_{VE} = 2$) and the defect (de, $c_{VE} \approx 1.98$) phases. $K_s^{(s)}(AB^{IIIa}, st)$ is much larger than $K_s^{(s)}(AB^{IIb})$. This is due to the occupied states at the Fermi level of the 5th band, which have a large s character (cf. Sect. C.IV). $K_s^{(s)}(AB^{IIIa}, de)$ is much smaller than $K_s^{(s)}(AB^{IIIa}, st)$ because the 5th band is unoccupied for the defect phases. For these phases the Fermi surface is formed by the 3rd and 4th bands which have a predominant p character at the Fermi level. The disagreement between $K_s^{(s)}(AB^{IIIa}, st)$ and $K_s^{exp.}$ supports the mechanism assumed for the defect structure in the AB^{IIIa} phases discussed in Sect. B[63], as this mechanism result in c_{VE} values <2 (see Table 2). Assuming a net paramagnetic orbital contribution $K_s^{(l)}$ (see below) and studying $K_s^{(s)}$ as a function of c_{VE} one finds that the agreement between K_s^{exp} and $K_s^{theor} \approx K_s^{(s)} + K_s^{(l)}$ is only satisfactory for $1.92 < c_{VE} < 1.98$[63]. Therefore, only the theoretical results for the defect phases are considered in the following discussion.

For the AB^{IIIa} compounds K_s of the alkali nuclei is almost zero. From the band structure it follows that this result is caused by two effects. Firstly, the density at the Fermi surface is very small (first term in Eqs. (36)) and secondly, the s-partial amplitudes for the alkali atomic spheres are very small for the electronic states at the Fermi level (cf. $q_s(E_F)/q(E_F)$ in Table 15). As the s partial density of states for the electronic states at the Fermi surface cannot be traced back to the total charge density distribution, the theoretical investigations show that it is not possible to deduce a charge distribution A^+B^- in Zintl phases from a vanishing Knight shift (cf. Sect. A).

For the non-alkali nuclei one finds the same trend as for the alkali nuclei. Again the Knight shifts $K_s(A\underline{B}^{IIIa})$ are small compared to the data for the pure metals because there are small s-partial densities for the electronic states at the Fermi surface. Except for NaTl the RAPW results of K_s given in Table 15 are smaller than the experimental data. Furthermore, the total spin contribution $K_s^{(\S)}$ is smaller than K_s^{dir}, although $K_s^{(\S)}$ includes the exchange enhancement effect (see above). For these phases it was discovered that the exchange enhancement effect is smaller than a negative exchange polarization effect[61, 63] caused by a negative spin density of the lower valence states (1st and 2nd bands in Fig. 5). For NaTl this negative contribution is even larger in size than the direct contact term, and $K_s^{\S}(N\underline{Tl})$ is negative, see Table 15.

K_s^{dir} was also calculated using the ASA method for LiAl, LiZn, and LiCd. The following values were found: $K_s(Li\underline{Al}) = 0.027\%$, $K_s(Li\underline{Zn}) = 0.16\%$ and $K_s(Li\underline{Cd}) = 0.27\%$. For LiAl this value is a factor of two larger than the RAPW and the experimental

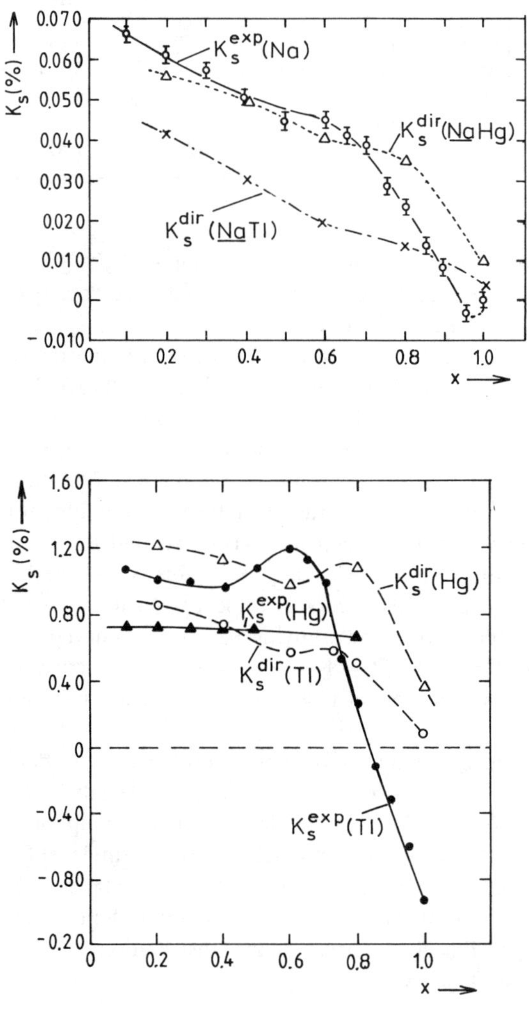

Fig. 20. Knight shift K_s of the Na-, Hg- and Tl-NMR in the system $NaHg_{1-x}Tl_x$ as a function of x. $K_s^{exp}(A)$ are the experimental results and $K_s^{dir}(A)$ and $K_s^{dir}(\underline{A}B)$ are the direct contribution for the A nuclei obtained from the band structures of NaTl and "NaHg(B32)" using the rigid band model (Schmidt et al.[18])

results, whereas for the AB compounds the RAPW results are larger than that of ASA. These trends in the differences between RAPW and ASA are the same as in the case of the density of states (cf. Table 7).

Let us next consider the orbital contribution $K_s^{(l)}$. This term is subdivided into a long-range diamagnetic term[97], a short-range diamagnetic term[98, 105] and a short-range paramagnetic term[106]. Based on a simple estimate one discovers for the Zintl phases considered here[63] that the absolute value of the total diamagnetic shielding should be small compared to $K_s^{(s)}$ and secondly, that the differences between K_s^{exp} and K_s can be explained by a paramagnetic orbital Knight shift[59, 63]. A numerical analysis to reproduce the

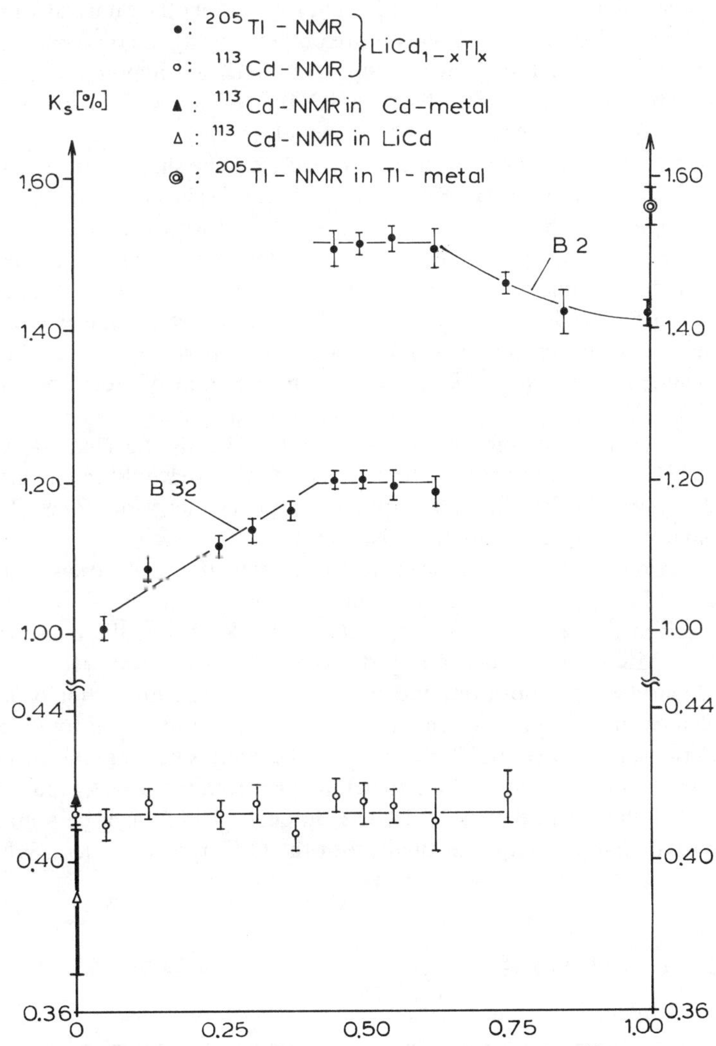

Fig. 21. Knight shift K_s of ^{205}Tl-NMR and ^{113}Cd-NMR in the alloys $LiCd_{1-x}Tl_x$ as a function of x (Baden et al.[17])

experimental results, however, cannot be given without adjustable parameters. Furthermore, for NaTl the experimental result K_s(Na<u>Tl</u>) is distinctly more negative than the theoretical value K_s(Na<u>Tl</u>). The origin of the large negative chemical shift of the Tl-NMR thus still requires an explanation.

Finally, in this section the Knight shift in two ternary systems $NaHg_{1-x}Tl_x$ and $LiCd_{1-x}Tl_x$ will be examined. The first system has been chosen for experimental and theoretical investigations in order to study changes in the unexplained diamagnetic shielding of the Tl nuclei, mentioned above, as a function of the valence electron concentration. Figure 20 shows the experimental and theoretical results for K_s in $NaHg_{1-x}Tl_x$. For this system the NaTl structure exists over nearly the whole range of composition (cf. Fig. 2). As a function of x, K_s(Tl) remains negative in the range of about $0.8 \leq x \leq 1.0$. For smaller values of x, the Tl Knight shift reflect the "normal" metallic (paramagnetic) behavior. The theoretical curves are obtained using the rigid band model (cf. Sect. C.VI) on the basis of the band structures of NaTl and hypothetical NaHg(B32). The band structure of both phases NaTl and NaHg(B32) are used to gather some information about the quality of the rigid band model for these systems. As only the direct part to K_s is calculated, the theoretical curves have the same shape as the s-partial densities of states within the energy range $8.5 < E_F/eV < 7.5$ displayed in Fig. 9.

Figure 20 shows that the decrease of K_s^{exp} with increasing x can be explained by a decrease of the s partial density of states for electronic states at the corresponding Fermi surface. Still unexplained is the origin of the large negative values of K_s^{exp}(Tl) for $x > 0.8$ (see also the discussion above on NaTl). For $x < 0.5$ the theoretical curve K_s^{dir}(Tl) is below the experimental curves because the paramagnetic orbital contribution is not included in the calculations. Seen from this point of view, the curve K_s^{dir}(Hg) lies too high. This curve, however, is deduced from the band structure of NaHg(B32) which does not take into account the interaction of the Hg and the Tl atoms. Guided by the experience from the study of Li_2CdIn (cf. Sect. C.IV) we would expect that the Tl atoms reduce the charge in the Hg atomic spheres and a smaller value for K_s(Hg) would result for a band structure calculation for Na_2HgTl.

In the second ternary system, $LiCd_{1-x}Tl_x$, the phase transition between the B2 and the B32 type of structure is studied (compare Fig. 3). From the experimental results given in Fig. 21 one notices that the Tl-NMR reflects the phase transition whereas the Cd-NMR is not influenced by it. As the Cd-NMR shift remains almost constant as a function of composition, the B2-B32 phase transition should not result in a drastic change in the s-partial density of states and with it in the total density of states at the Fermi surface. For the Tl-NMR the Knight shift is larger in the B2 phases than in the B32 phases. This appears to be understandable from the investigation of the band structures of LiTl and hypothetical LiTl(B32) in Sect. D.I. As can be seen from Table 12, the s partial charge is larger in the case of the CsCl type of structure than for NaTl.

F. Conclusions

In the present work electronic properties and the nature of chemical bonding in intermetallic B32-type Zintl phases are discussed on the basis of relativistic and non-relativistic as well as spin polarized band structure calculations.

The theoretical investigations give some insight into the covalent, metallic, and ionic contributions to the chemical bonds and show the charge transfer from the alkali to the non-alkali sublattices. Furthermore, the role of the sizes of the atoms on the stability of the B32 type Zintl phases is analyzed. A comparison of the electronic states in B2 and B32 type phases informs us about the different mechanisms which have to be taken into consideration for studying B2-B32 phase transitions. The optical absorption, magnetic susceptibility, and Knight shift in the B32 type Zintl phases deviate from the behaviour of most intermetallic phases. The data have been interpreted satisfactorily on the basis of the shapes of the electronic bands, the density of states, and the charge analysis.

Acknowledgements. The author wishes to thank Professor Dr. Helmut Witte for many helpful discussions and Professor Dr. Alarich Weiss for continuous support and interest in the present work. I am also grateful to the Fonds der Chemischen Industrie for support of this work.

G. References

1. Schäfer, H., Eisenmann, B., Müller, W.: Angew. Chem. *85,* 742 (1973); Angew. Chem. Int. Ed. Engl. *12,* 694 (1973)
2. Laves, F.: Naturwissenschaften *29,* 244 (1941)
3. See e.g.: Klemm, W.: Proc. Chem. Soc. London (1958), p. 329
   Von Schnering, H. G.: Angew. Chem. *93,* 44 (1981); Angew. Chem. Int. Ed. Engl. *20,* 33 (1981)
   Schäfer, H.: Ann. Rev. Mater. Sci. *15,* 1 (1985)
4. Zintl, E., Brauer, G.: Z. Phys. Chem. B *20,* 245 (1933)
   Zintl, E., Dullenkopf, W.: ibid. *16,* 195 (1932)
   Zintl, E., Neumayr, S.: ibid. *20,* 272 (1933)
   Zintl, E., Schneider, A.: Z. Elektrochem. *41,* 294 (1935)
5. Zintl, E., Woltersdorf, G.: ibid. *41,* 876 (1935)
6. Komowsky, G., Maximow, A.: Z. Kristallogr. *92,* 274 (1935)
7. Schneider, A., Heymer, G.: Z. Anorg. Allgem. Chem. *286,* 118 (1956)
8. Levine, E. D., Rapperport, E. J.: Trans. AIME *227,* 1204 (1963)
9. Eckerlin, P., Maak, I., Rabenau, A.: Z. Anorg. Allgem. Chem. *327,* 143 (1964)
10. Pauly, H., Weiss, Al., Witte, H.: Z. Metallkd. *59,* 47 (1968)
11. Pauly, H., Weiss, Al., Witte, H.: ibid. *59,* 554 (1968)
12. Abdulahad, I., Zwilling, M., Weiss, Al.: ibid. *62,* 231 (1971)
13. Bartsch, R., Weiss, Al.: ibid. *66,* 39 (1975)
14. Kuriyama, K., Masaki, N.: Acta Crystallogr. B *31,* 1793 (1975)
15. Yahagi, M., Kuriyama, K., Iwamura, K.: Jpn. J. Appl. Phys. *14,* 405 (1975)
16. Kishio, K., Brittain, J. O.: J. Phys. Chem. Solids *40,* 933 (1979)
17. Baden, W., Schmidt, P. C., Weiss, Al.: phys. stat. sol. (a) *51,* 183 (1979)
18. Schmidt, P. C., Baden, W., Weiden, N., Weiss, Al.: ibid. *92,* 205 (1985)
19. Grube, G., Schmidt, A.: Z. Elektrochem. *42,* 201 (1936)
20. Thummel, R., Klemm, W.: Z. Anorg. Allg. Chem. *376,* 44 (1970)
21. Grube, G., Voßkühler, H., Vogt, H.: Z. Elektrochem. *38,* 869 (1932)
22. Kuriyama, K., Kamijoh, T., Nozaki, T.: Phys. Rev. B *21,* 4887 (1980)
23. Junod, P., Mooser, E.: Helv. Phys. Acta *29,* 194 (1956)
24. Kuriyama, K., Yahagi, M., Nakaya, T., Ashida, T., Satoh, K., Iwamura, K.: Phys. Rev. B *29,* 5924 (1984)
25. Klemm, W., Fricke, H.: Z. Anorg. Allg. Chem. *282,* 162 (1955)
26. Yao, Y. L.: Trans. Soc. AIME *230,* 1725 (1964)
27. Bloembergen, N., Rowland, T. J.: Acta Met. *1,* 731 (1953)

28. Schone, H. E., Knight, W. D.: ibid. *11,* 179 (1963)
29. Bennett, L. H.: ibid. *14,* 997 (1966)
30. Bennett, L. H.: Bull. Am. Phys. Soc. *11,* 172 (1966)
31. Bennett, L. H.: Phys. Rev. *150,* 418 (1966)
32. Carter, G. C., Bennett, L. H., Kahan, D. J.: Metallic Shifts in NMR, in: Chalmers, B., Christian, J. W., Massalski, T. B. (Ed.), Progress in Material Science, Vol. 20, Pergamon Press, New York (1977)
33. Kuriyama, K., Saito, S.: Phys. Rev. B *13,* 1528 (1976)
34. Kuriyama, K., Saito, S., Iwamura, K.: J. Phys. Chem. Sol. *40,* 457 (1979)
35. Zwilling, M., Schmidt, P. C., Weiss, Al.: J. Appl. Phys. *16,* 255 (1978)
36. Ratajack, M. T., Kisho, K., Brittain, J. O., Kannewurf, C. R.: Phys. Rev. B *21,* 2144 (1980)
37. Yao, N. P., Heredy, L. A., Saunders, R. C.: J. Electrochem. Soc. *118,* 1039 (1971)
38. Melendres, C. A., Sy, C. C.: ibid. *125,* 727 (1978)
39. Wen, C. John, Bonkamp, B. A., Huggins, R. A., Weppner, W.: ibid. *126,* 2258 (1979)
40. Anderson, P. W.: Concepts in Solids, Benjamin, N.Y. (1963), p. 6
41. Brewer, L.: High Strength Materials, Ed. V. Zackay, Wiley, N.Y. (1964), p. 37
42. Engel, N. N.: Developments in the Structural Chemistry of Alloy Phases, Ed. B. Giessen, Plenum Press, N.Y. (1969), pp. 25–40
43. Pearson, W. B.: Developments in the Structural Chemistry of Alloy Phases, Ed. B. Giessen, Plenum Press, N.Y. (1969), pp. 45–48
44. Hückel, W.: Structural Chemistry of Inorganic Compounds, Elsevier, Amsterdam (1951), pp. 829ff.
45. Hansen, M.: Constitution of Binary Alloys, McGraw-Hill, New York (1958)
46. Klemm, W.: Ref. 3
47. Watson, R. E., Bennett, L. H., Carter, G. C., Weisman, I. D.: Phys. Rev. B *3,* 222 (1971)
48. Zintl, E.: Angew. Chem. *52,* 1 (1939)
49. Mooser, E., Pearson, W. B.: Phys. Rev. *101,* 1608 (1956)
50. Klemm, W., Busmann, E.: Z. Anorg. Allg. Chem. *319,* 297 (1963)
51. Schäfer, H., Eisenmann, B.: Rev. Inorg. Chem. *3,* 29 (1981)
52. Pearson, W. B.: Acta Crystallogr. *17,* 1 (1964)
53. von Schnering, H. G.: Ref. 3
54. Inglesfield, J. E.: J. Phys. *C4,* 1003 (1971)
55. McNeil, M. B., Pearson, W. B., Bennett, L. H., Watson, R. E.: ibid. *C6,* 1 (1973)
56. Makhnovetskii, A. B., Krasko, G. L.: phys. stat. sol. (b) *80,* 341 (1977)
57. Zunger, A.: Phys. Rev. B *17,* 2582 (1978)
58. Asada, T., Jarlborg, T., Freeman, A. J.: ibid. *24,* 510 (1981)
59. Asada, T., Jarlborg, T., Freeman, A. J.: ibid. *24,* 857 (1981)
60. Robertson, J.: Sol. State Com. *47,* 899 (1983)
61. Schmidt, P. C.: Phys. Rev. B *31,* 5015 (1985)
62. Schmidt, P. C.: Z. Naturf. *40a,* 335 (1985)
63. Schmidt, P. C.: Phys. Rev. B *34,* 3868 (1986)
64. Christensen, N. E.: ibid. *32,* 207 (1985)
65. Schwarz, K., Mohn, P., Blaha, P., Kübler, J.: J. Phys. F *14,* 2659 (1984)
66. Laves, F.: Theory of Alloy Phases (Amer. Soc. Met., Cleveland, Ohio, 1956), p. 124
67. Goldschmidt, V. M.: Z. Phys. Chem. *133,* 396 (1928)
68. Neuburger, M. C.: Z. Krist. *93,* 1 (1936)
69. Hume-Rothery, W., Raynor, G. V.: The Structure of Metals and Alloys, Institute of Metals, London, pp. 87–89 (1954)
70. Huggins, M. L.: Phys. Rev. *28,* 1086 (1926)
71. Pauling, L., Huggins, M. L.: Z. Krist. *87,* 205 (1934)
72. Kishio, K., Owers-Bradley, J. R., Halperin, W. P., Brittain, J. O.: J. Phys. Chem. Solids *42,* 1031 (1981)
73. Loucks, T.: Augmented Plane Wave Method, Benjamin, N.Y. (1967)
74. Dimmock, J. O.: Solid State Physics *26,* 104 (1971)
75. Rose, E. M.: Relativistic Electron Theory, John Wiley, N.Y. (1961)
76. Hohenberg, P., Kohn, W.: Phys. Rev. *136,* B864 (1964)
77. Kohn, W., Sham, L. J.: ibid. *140,* A1133 (1965)
78. MacDonald, A. H., Vosko, S. H.: J. Phys. C *12,* 2977 (1979)

79. Vosko, S. H., Wilk, L., Nusair, M.: Can. J. Phys. *58,* 1200 (1980)
80. Andersen, O. K.: Phys. Rev. Lett. *27,* 1211 (1971)
81. Andersen, O. K.: Phys. Rev. B *12,* 3060 (1975)
82. Friedel, J.: Adv. Phys. *3,* 446 (1954)
83. Kutzelnigg, W.: Angew. Chem. *96,* 262 (1984); Angew. Chem. Int. Ed. Engl. *23,* 272 (1984)
84. Wiberg, K. B.: J. Am. Soc. *102,* 1229 (1980)
85. Bader, R. F. W.: Acc. Chem. Res. *18,* 9 (1985)
Bader, R. F. W., Nguyen-Dang, T. T.: Adv. Qu. Chem. *14,* 63 (1981)
86. Schmidt, P. C.: Ber. Bunsenges. Phys. Chem. *79,* 1064 (1975)
87. Scheuring, T., Weil, K. G.: Int J. Mass Spectr. and Ion Phys. *47,* 227 (1983)
88. Slater, J. C.: J. Chem. Phys. *57,* 2389 (1972)
89. Janak, J. F.: Phys. Rev. B *9,* 3985 (1974)
90. Pettifor, D. G.: J. Phys. C *3,* 367 (1970)
91. Heine, V.: Solid State Physics *35,* 1 (1980)
92. Andersen, O. K., Christensen, N. E.: to be published
93. Wooten, F.: Optical Properties of Solids, Academic Press, New York, 1972, pp. 101–116
94. Hebborn, J. E., Sondheimer, E. H.: J. Phys. Chem. Solids *13,* 105 (1960)
95. Roth, L. M.: ibid. *23,* 433 (1962)
96. Blount, E. I.: Phys. Rev. *126,* 1636 (1962)
97. Das, T. P., Sondheimer, E. H.: Phil. Mag. *5,* 529 (1960)
98. Hebborn, J. E.: Proc. Phys. Soc. *80,* 1237 (1962)
99. Pauli, W.: Z. Physik *41,* 81 (1927)
100. Mahanti, S. D., Das, T. P.: Phys. Rev. B *3,* 1599 (1971)
101. von Barth, U., Hedin, L.: J. Phys. C *5,* 1629 (1972)
102. Knight, W. D.: Solid State Physics *2,* 93 (1956)
103. Cohen, M. H., Goldings, D. A., Heine, V.: Proc. Phys. Soc. London *73,* 811 (1959)
104. Gaspari, G. D., Shyn, W. M., Das, T. P.: Phys. Rev. *134,* A852 (1964)
105. Lamb, W.: ibid. *60,* 817 (1941)
106. Yafet, Y., Jaccarino, V.: ibid. *133,* A1630 (1964)
Switendick, J. A., Gossard, A. C., Jaccarino, V.: ibid. *136,* A1119 (1964)

Author-Index Volumes 1–65

Ahrland, S.: Factors Contributing to (b)-behaviour in Acceptors. Vol. 1, pp. 207–220.

Ahrland, S.: Thermodynamics of Complex Formation between Hard and Soft Acceptors and Donors. Vol. 5, pp. 118–149.

Ahrland, S.: Thermodynamics of the Stepwise Formation of Metal-Ion Complexes in Aqueous Solution. Vol. 15, pp. 167–188.

Allen, G. C., Warren, K. D.: The Electronic Spectra of the Hexafluoro Complexes of the First Transition Series. Vol. 9, pp. 49–138.

Allen, G. C., Warren, K. D.: The Electronic Spectra of the Hexafluoro Complexes of the Second and Third Transition Series. Vol. 19, pp. 105–165.

Ardon, M., Bino, A.: A New Aspect of Hydrolysis of Metal Ions: The Hydrogen-Oxide Bridging Ligand ($H_3O_2^-$). Vol. 65, pp. 1–28.

Averill, B. A.: Fe–S and Mo–Fe–S Clusters as Models for the Active Site of Nitrogenase. Vol. 53, pp. 57–101.

Babel, D.: Structural Chemistry of Octahedral Fluorocomplexes of the Transition Elements. Vol. 3, pp. 1–87.

Bacci, M.: The Role of Vibronic Coupling in the Interpretation of Spectroscopic and Structural Properties of Biomolecules. Vol. 55, pp. 67–99.

Baker, E. C., Halstead, G.W., Raymond, K. N.: The Structure and Bonding of 4*f* and 5*f* Series Organometallic Compounds. Vol. 25, pp. 21–66.

Balsenc, L. R.: Sulfur Interaction with Surfaces and Interfaces Studied by Auger Electron Spectrometry. Vol. 39, pp. 83–114.

Banci, L., Bencini, A., Benelli, C., Gatteschi, D., Zanchini, C.: Spectral-Structural Correlations in High-Spin Cobalt(II) Complexes. Vol. 52, pp. 37–86.

Baughan, E. C.: Structural Radii, Electron-cloud Radii, Ionic Radii and Solvation. Vol. 15, pp. 53–71.

Bayer, E., Schretzmann, P.: Reversible Oxygenierung von Metallkomplexen. Vol. 2, pp. 181–250.

Bearden, A. J., Dunham, W. R.: Iron Electronic Configurations in Proteins: Studies by Mössbauer Spectroscopy. Vol. 8, pp. 1–52.

Bertini, I., Luchinat, C., Scozzafava, A.: Carbonic Anhydrase: An Insight into the Zinc Binding Site and into the Active Cavity Through Metal Substitution. Vol. 48, pp. 45–91.

Blasse, G.: The Influence of Charge-Transfer and Rydberg States on the Luminescence Properties of Lanthanides and Actinides. Vol. 26, pp. 43–79.

Blasse, G.: The Luminescence of Closed-Shell Transition Metal-Complexes. New Developments. Vol. 42, pp. 1–41.

Blauer, G.: Optical Activity of Conjugated Proteins. Vol. 18, pp. 69–129.

Bleijenberg, K. C.: Luminescence Properties of Uranate Centres in Solids. Vol. 42, pp. 97–128.

Boeyens, J. C. A.: Molecular Mechanics and the Structure Hypothesis. Vol. 63, pp. 65–101.

Bonnelle, C.: Band and Localized States in Metallic Thorium, Uranium and Plutonium, and in Some Compounds, Studied by X-Ray Spectroscopy. Vol. 31, pp. 23–48.

Bradshaw, A. M., Cederbaum, L. S., Domcke, W.: Ultraviolet Photoelectron Spectroscopy of Gases Adsorbed on Metal Surfaces. Vol. 24, pp. 133–170.

Braterman, P. S.: Spectra and Bonding in Metal Carbonyls. Part A: Bonding. Vol. 10, pp. 57–86.

Braterman, P. S.: Spectra and Bonding in Metal Carbonyls. Part B: Spectra and Their Interpretation. Vol. 26, pp. 1–42.

Bray, R. C., Swann, J. C.: Molybdenum-Containing Enzymes. Vol. 11, pp. 107–144.

Brooks, M. S. S.: The Theory of 5f Bonding in Actinide Solids. Vol. 59/60, pp. 263–293.

van Bronswyk, W.: The Application of Nuclear Quadrupole Resonance Spectroscopy to the Study of Transition Metal Compounds. Vol. 7, pp. 87–113.

Buchanan, B. B.: The Chemistry and Function of Ferredoxin. Vol. 1, pp. 109–148.

Buchler, J. W., Kokisch, W., Smith, P. D.: *Cis, Trans,* and *Metal* Effects in Transition Metal Porphyrins. Vol. 34, pp. 79–134.

Bulman, R. A.: Chemistry of Plutonium and the Transuranics in the Biosphere. Vol. 34, pp. 39–77.

Burdett, J. K.: The Shapes of Main-Group Molecules; A Simple Semi Quantitative Molecular Orbital Approach. Vol. 31, pp. 67–105.

Burdett, J. K.: Some Structural Problems Examined Using the Method of Moments. Vol. 65, pp. 29–90.

Campagna, M., Wertheim, G. K., Bucher, E.: Spectroscopy of Homogeneous Mixed Valence Rare Earth Compounds. Vol. 30, pp. 99–140.

Chasteen, N. D.: The Biochemistry of Vanadium, Vol. 53, pp. 103–136.

Cheh, A. M., Neilands, J. P.: The δ-Aminolevulinate Dehydratases: Molecular and Environmental Properties. Vol. 29, pp. 123–169.

Ciampolini, M.: Spectra of 3 d Five-Coordinate Complexes. Vol. 6, pp. 52–93.

Chimiak, A., Neilands, J. B.: Lysine Analogues of Siderophores. Vol. 58, pp. 89–96.

Clack, D. W., Warren, K. D.: Metal-Ligand Bonding in 3d Sandwich Complexes, Vol. 39, pp. 1–41.

Clark, R. J. H., Stewart, B.: The Resonance Raman Effect. Review of the Theory and of Applications in Inorganic Chemistry. Vol. 36, pp. 1–80.

Clarke, M. J., Fackler, P. H.: The Chemistry of Technetium: Toward Improved Diagnostic Agents. Vol. 50, pp. 57–78.

Cohen, I. A.: Metal-Metal Interactions in Metalloporphyrins, Metalloproteins and Metalloenzymes. Vol. 40, pp. 1–37.

Connett, P. H., Wetterhahn, K. E.: Metabolism of the Carcinogen Chromate by Cellular Constitutents. Vol. 54, pp. 93–124.

Cook, D. B.: The Approximate Calculation of Molecular Electronic Structures as a Theory of Valence. Vol. 35, pp. 37–86.

Cotton, F. A., Walton, R. A.: Metal-Metal Multiple Bonds in Dinuclear Clusters. Vol. 62, pp. 1–49.

Cox, P. A.: Fractional Parentage Methods for Ionisation of Open Shells of *d* and *f* Electrons. Vol. 24, pp. 59–81.

Crichton, R. R.: Ferritin. Vol. 17, pp. 67–134.

Daul, C., Schläpfer, C. W., von Zelewsky, A.: The Electronic Structure of Cobalt(II) Complexes with Schiff Bases and Related Ligands. Vol. 36, pp. 129–171.

Dehnicke, K., Shihada, A.-F.: Structural and Bonding Aspects in Phosphorus Chemistry-Inorganic Derivates of Oxohalogeno Phosphoric Acids. Vol. 28, pp. 51–82.

Dobiáš, B.: Surfactant Adsorption on Minerals Related to Flotation. Vol. 56, pp. 91–147.

Doughty, M. J., Diehn, B.: Flavins as Photoreceptor Pigments for Behavioral Responses. Vol. 41, pp. 45–70.

Drago, R. S.: Quantitative Evaluation and Prediction of Donor-Acceptor Interactions. Vol. 15, pp. 73–139.

Duffy, J. A.: Optical Electronegativity and Nephelauxetic Effect in Oxide Systems. Vol. 32, pp. 147–166.

Dunn, M. F.: Mechanisms of Zinc Ion Catalysis in Small Molecules and Enzymes. Vol. 23, pp. 61–122.

Emsley, E.: The Composition, Structure and Hydrogen Bonding of the β-Deketones. Vol. 57, pp. 147–191.

Englman, R.: Vibrations in Interaction with Impurities. Vol. 43, pp. 113–158.

Epstein, I. R., Kustin, K.: Design of Inorganic Chemical Oscillators. Vol. 56, pp. 1–33.

Ermer, O.: Calculations of Molecular Properties Using Force Fields. Applications in Organic Chemistry. Vol. 27, pp. 161–211.

Ernst, R. D.: Structure and Bonding in Metal-Pentadienyl and Related Compounds. Vol. 57, pp. 1–53.

Erskine, R. W., Field, B. O.: Reversible Oxygenation. Vol. 28, pp. 1–50.

Fajans, K.: Degrees of Polarity and Mutual Polarization of Ions in the Molecules of Alkali Fluorides, SrO, and BaO. Vol. 3, pp. 88–105.

Fee, J. A.: Copper Proteins – Systems Containing the "Blue" Copper Center. Vol. 23, pp. 1–60.

Feeney, R. E., Komatsu, S. K.: The Transferrins. Vol. 1, pp. 149–206.

Felsche, J.: The Crystal Chemistry of the Rare-Earth Silicates. Vol. 13, pp. 99–197.

Ferreira, R.: Paradoxical Violations of Koopmans' Theorem, with Special Reference to the 3 d Transition Elements and the Lanthanides. Vol. 31, pp. 1–21.

Fidelis, I. K., Mioduski, T.: Double-Double Effect in the Inner Transition Elements. Vol. 47, pp. 27–51.

Fournier, J. M.: Magnetic Properties of Actinide Solids. Vol. 59/60, pp. 127–196.

Fournier, J. M., Manes, L.: Actinide Solids. 5 f Dependence of Physical Properties. Vol. 59/60, pp. 1–56.

Fraga, S., Valdemoro, C.: Quantum Chemical Studies on the Submolecular Structure of the Nucleic Acids. Vol. 4, pp. 1–62.

Fraústo da Silva, J. J. R., Williams, R. J. P.: The Uptake of Elements by Biological Systems. Vol. 29, pp. 67–121.

Fricke, B.: Superheavy Elements. Vol. 21, pp. 89–144.

Fuhrhop, J.-H.: The Oxidation States and Reversible Redox Reactions of Metalloporphyrins. Vol. 18, pp. 1–67.

Furlani, C., Cauletti, C.: He(I) Photoelectron Spectra of *d*-metal Compounds. Vol. 35, pp. 119–169.

Gerloch, M., Harding, J. H., Woolley, R. G.: The Context and Application of Ligand Field Theory. Vol. 46, pp. 1–46.

Gillard, R. D., Mitchell, P. R.: The Absolute Configuration of Transition Metal Complexes. Vol. 7, pp. 46–86.

Gleitzer, C., Goodenough, J. B.: Mixed-Valence Iron Oxides. Vol. 61, pp. 1–76.

Gliemann, G., Yersin, H.: Spectroscopic Properties of the Quasi One-Dimensional Tetracyanoplatinate(II) Compounds. Vol. 62, pp. 87–153.

Golovina, A. P., Zorov, N. B., Runov, V. K.: Chemical Luminescence Analysis of Inorganic Substances. Vol. 47, pp. 53–119.

Green, J. C.: Gas Phase Photoelectron Spectra of *d*- and *f*-Block Organometallic Compounds. Vol. 43, pp. 37–112.

Grenier, J. C., Pouchard, M., Hagenmuller, P.: Vacancy Ordering in Oxygen-Deficient Perovskite-Related Ferrities. Vol. 47, pp. 1–25.

Griffith, J. S.: On the General Theory of Magnetic Susceptibilities of Polynuclear Transitionmetal Compounds. Vol. 10, pp. 87–126.

Gubelmann, M. H., Williams, A. F.: The Structure and Reactivity of Dioxygen Complexes of the Transition Metals. Vol. 55, pp. 1–65.

Guilard, R., Lecomte, C., Kadish, K. M.: Synthesis, Electrochemistry, and Structural Properties of Porphyrins with Metal-Carbon Single Bonds and Metal-Metal Bonds. Vol. 64, pp. 205–268.

Gütlich, P.: Spin Crossover in Iron(II)-Complexes. Vol. 44, pp. 83–195.

Gutmann, V., Mayer, U.: Thermochemistry of the Chemical Bond. Vol. 10, pp. 127–151.

Gutmann, V., Mayer, U.: Redox Properties: Changes Effected by Coordination. Vol. 15, pp. 141–166.

Gutmann, V., Mayer, H.: Application of the Functional Approach to Bond Variations under Pressure. Vol. 31, pp. 49–66.

Hall, D. I., Ling, J. H., Nyholm, R. S.: Metal Complexes of Chelating Olefin-Group V Ligands. Vol. 15, pp. 3–51.

Harnung, S. E., Schäffer, C. E.: Phase-fixed 3-Γ Symbols and Coupling Coefficients for the Point Groups. Vol. 12, pp. 201–255.

Harnung, S. E., Schäffer, C. E.: Real Irreducible Tensorial Sets and their Application to the Ligand-Field Theory. Vol. 12, pp. 257–295.

Hathaway, B. J.: The Evidence for "Out-of-the-Plane" Bonding in Axial Complexes of the Copper(II) Ion. Vol. 14, pp. 49–67.

Hathaway, B. J.: A New Look at the Stereochemistry and Electronic Properties of Complexes of the Copper(II) Ion. Vol. 57, pp. 55–118.

Hellner, E. E.: The Frameworks (Bauverbände) of the Cubic Structure Types. Vol. 37, pp. 61–140.

von Herigonte, P.: Electron Correlation in the Seventies. Vol. 12, pp. 1–47.

Hemmerich, P., Michel, H., Schug, C., Massey, V.: Scope and Limitation of Single Electron Transfer in Biology. Vol. 48, pp. 93–124.

Hider, R. C.: Siderophores Mediated Absorption of Iron. Vol. 58, pp. 25–88.

Hill, H. A. O., Röder, A., Williams, R. J. P.: The Chemical Nature and Reactivity of Cytochrome P-450. Vol. 8, pp. 123–151.

Hogenkamp, H. P. C., Sando, G. N.: The Enzymatic Reduction of Ribonucleotides. Vol. 20, pp. 23–58.

Hoffmann, D. K., Ruedenberg, K., Verkade, J. G.: Molecular Orbital Bonding Concepts in Polyatomic Molecules – A Novel Pictorial Approach. Vol. 33, pp. 57–96.

Hubert, S., Hussonnois, M., Guillaumont, R.: Measurement of Complexing Constants by Radiochemical Methods. Vol. 34, pp. 1–18.

Hudson, R. F.: Displacement Reactions and the Concept of Soft and Hard Acids and Bases. Vol. 1, pp. 221–223.

Hulliger, F.: Crystal Chemistry of Chalcogenides and Pnictides of the Transition Elements. Vol. 4, pp. 83–229.

Ibers, J. A., Pace, L. J., Martinsen, J., Hoffman, B. M.: Stacked Metal Complexes: Structures and Properties. Vol. 50, pp. 1–55.

Iqbal, Z.: Intra- und Inter-Molecular Bonding and Structure of Inorganic Pseudohalides with Triatomic Groupings. Vol. 10, pp. 25–55.

Izatt, R. M., Eatough, D. J., Christensen, J. J.: Thermodynamics of Cation-Macrocyclic Compound Interaction. Vol. 16, pp. 161–189.

Jain, V. K., Bohra, R., Mehrotra, R. C.: Structure and Bonding in Organic Derivatives of Antimony(V). Vol. 52, pp. 147–196.

Jerome-Lerutte, S.: Vibrational Spectra and Structural Properties of Complex Tetracyanides of Platinum, Palladium and Nickel. Vol. 10, pp. 153–166.

Jørgensen, C. K.: Electric Polarizability, Innocent Ligands and Spectroscopic Oxidation States. Vol. 1, pp. 234–248.

Jørgensen, C. K.: Recent Progress in Ligand Field Theory. Vol. 1, pp. 3–31.

Jørgensen, C. K.: Relations between Softness, Covalent Bonding, Ionicity and Electric Polarizability. Vol. 3, pp. 106–115.

Jørgensen, C. K.: Valence-Shell Expansion Studied by Ultra-violet Spectroscopy. Vol. 6, pp. 94–115.

Jørgensen, C. K.: The Inner Mechanism of Rare Earths Elucidated by Photo-Electron Spectra. Vol. 13, pp. 199–253.

Jørgensen, C. K.: Partly Filled Shells Constituting Anti-bonding Orbitals with Higher Ionization Energy than their Bonding Counterparts. Vol. 22, pp. 49–81.

Jørgensen, C. K.: Photo-electron Spectra of Non-metallic Solids and Consequences for Quantum Chemistry. Vol. 24, pp. 1–58.

Jørgensen, C. K.: Narrow Band Thermoluminescence (Candoluminescence) of Rare Earths in Auer Mantles. Vol. 25, pp. 1–20.

Jørgensen, C. K.: Deep-lying Valence Orbitals and Problems of Degeneracy and Intensities in Photoelectron Spectra. Vol. 30, pp. 141–192.

Jørgensen, C. K.: Predictable Quarkonium Chemistry. Vol. 34, pp. 19–38.

Jørgensen, C. K.: The Conditions for Total Symmetry Stabilizing Molecules, Atoms, Nuclei and Hadrons. Vol. 43, pp. 1–36.

Jørgensen, C. K., Reisfeld, R.: Uranyl Photophysics. Vol. 50, pp. 121–171.

O'Keeffe, M., Hyde, B. G.: An Alternative Approach to Non-Molecular Crystal Structures with Emphasis on the Arrangements of Cations. Vol. 61, pp. 77–144.

Kimura, T.: Biochemical Aspects of Iron Sulfur Linkage in None-Heme Iron Protein, with Special Reference to "Adrenodoxin". Vol. 5, pp. 1–40.

Kitagawa, T., Ozaki, Y.: Infrared and Raman Spectra of Metalloporphyrins. Vol. 64, pp. 71–114.

Kiwi, J., Kalyanasundaram, K., Grätzel, M.: Visible Light Induced Cleavage of Water into Hydrogen and Oxygen in Colloidal and Microheterogeneous Systems. Vol. 49, pp. 37–125.

Kjekshus, A., Rakke, T.: Considerations on the Valence Concept. Vol. 19, pp. 45–83.

Kjekshus, A., Rakke, T.: Geometrical Considerations on the Marcasite Type Structure. Vol. 19, pp. 85–104.

König, E.: The Nephelauxetic Effect. Calculation and Accuracy of the Interelectronic Repulsion Parameters I. Cubic High-Spin d^2, d^3, d^7 and d^8 Systems. Vol. 9, pp. 175–212.

Koppikar, D. K., Sivapullaiah, P. V., Ramakrishnan, L., Soundararajan, S.: Complexes of the Lanthanides with Neutral Oxygen Donor Ligands. Vol. 34, pp. 135–213.

Krumholz, P.: Iron(II) Diimine and Related Complexes. Vol. 9, pp. 139–174.

Kustin, K., McLeod, G. C., Gilbert, T. R., Briggs, LeB. R., 4th.: Vanadium and Other Metal Ions in the Physiological Ecology of Marine Organisms. Vol. 53, pp. 137–158.

Labarre, J. F.: Conformational Analysis in Inorganic Chemistry: Semi-Empirical Quantum Calculation vs. Experiment. Vol. 35, pp. 1–35.

Lammers, M., Follmann, H.: The Ribonucleotide Reductases: A Unique Group of Metalloenzymes Essential for Cell Proliferation. Vol. 54, pp. 27–91.

Lehn, J.-M.: Design of Organic Complexing Agents. Strategies towards Properties. Vol. 16, pp. 1–69.

Linarès, C., Louat, A., Blanchard, M.: Rare-Earth Oxygen Bonding in the $LnMO_4$Xenotime Structure. Vol. 33, pp. 179–207.

Lindskog, S.: Cobalt(II) in Metalloenzymes. A Reporter of Structure-Function Relations. Vol. 8, pp. 153–196.

Liu, A., Neilands, J. B.: Mutational Analysis of Rhodotorulic Acid Synthesis in *Rhodotorula pilimanae*. Vol. 58, pp. 97–106.

Livorness, J., Smith, T.: The Role of Manganese in Photosynthesis. Vol. 48, pp. 1–44.

Llinás, M.: Metal-Polypeptide Interactions: The Conformational State of Iron Proteins. Vol. 17, pp. 135–220.

Lucken, E. A. C.: Valence-Shell Expansion Studied by Radio-Frequency Spectroscopy. Vol. 6, pp. 1–29.

Ludi, A., Güdel, H. U.: Structural Chemistry of Polynuclear Transition Metal Cyanides. Vol. 14, pp. 1–21.

Maggiora, G. M., Ingraham, L. L.: Chlorophyll Triplet States. Vol. 2, pp. 126–159.

Magyar, B.: Salzebullioskopie III. Vol. 14, pp. 111–140.

Makovicky, E., Hyde, B. G.: Non-Commensurate (Misfit) Layer Structures. Vol. 46, pp. 101–170.

Manes, L., Benedict, U.: Structural and Thermodynamic Properties of Actinide Solids and Their Relation to Bonding. Vol. 59/60, pp. 75–125.

Mann, S.: Mineralization in Biological Systems. Vol. 54, pp. 125–174.

Mason, S. F.: The Ligand Polarization Model for the Spectra of Metal Complexes: The Dynamic Coupling Transition Probabilities. Vol. 39, pp. 43–81.

Mathey, F., Fischer, J., Nelson, J. H.: Complexing Modes of the Phosphole Moiety. Vol. 55, pp. 153–201.

Mayer, U., Gutmann, V.: Phenomenological Approach to Cation-Solvent Interactions. Vol. 12, pp. 113–140.

Mildvan, A. S., Grisham, C. M.: The Role of Divalent Cations in the Mechanism of Enzyme Catalyzed Phosphoryl and Nucleotidyl. Vol. 20, pp. 1–21.

Mingos, D. M. P., Hawes, J. C.: Complementary Spherical Electron Density Model. Vol. 63, pp. 1–63.

Moreau-Colin, M. L.: Electronic Spectra and Structural Properties of Complex Tetracyanides of Platinum, Palladium and Nickel. Vol. 10, pp. 167–190.

Morgan, B., Dolphin, D.: Synthesis and Structure of Biometic Porphyrins. Vol. 64, pp. 115–204.

Morris, D. F. C.: Ionic Radii and Enthalpies of Hydration of Ions. Vol. 4, pp. 63–82.

Morris, D. F. C.: An Appendix to Structure and Bonding. Vol. 4 (1968). Vol. 6, pp. 157–159.

Müller, A., Baran, E. J., Carter, R. O.: Vibrational Spectra of Oxo-, Thio-, and Selenometallates of Transition Elements in the Solid State. Vol. 26, pp. 81–139.

Müller, A., Diemann, E., Jørgensen, C. K.: Electronic Spectra of Tetrahedral Oxo, Thio and Seleno Complexes Formed by Elements of the Beginning of the Transition Groups. Vol. 14, pp. 23–47.

Müller, U.: Strukturchemie der Azide. Vol. 14, pp. 141–172.

Müller, W., Spirlet, J.-C.: The Preparation of High Purity Actinide Metals and Compounds. Vol. 59/60, pp. 57–73.

Murrell, J. N.: The Potential Energy Surfaces of Polyatomic Molecules. Vol. 32, pp. 93–146.

Naegele, J. R., Ghijsen, J.: Localization and Hybridization of 5f States in the Metallic and Ionic Bond as Investigated by Photoelectron Spectroscopy. Vol. 59/60, pp. 197–262.

Nag, K., Bose, S. N.: Chemistry of Tetra- and Pentavalent Chromium. Vol. 63, pp. 153–197.

Neilands, J. B.: Naturally Occurring Non-porphyrin Iron Compounds. Vol. 1, pp. 59–108.

Neilands, J. B.: Evolution of Biological Iron Binding Centers. Vol. 11, pp. 145–170.

Neilands, J. B.: Methodology of Siderophores. Vol. 58, pp. 1–24.

Nieboer, E.: The Lanthanide Ions as Structural Probes in Biological and Model Systems. Vol. 22, pp. 1–47.

Novack, A.: Hydrogen Bonding in Solids. Correlation of Spectroscopic and Christallographic Data. Vol. 18, pp. 177–216.

Nultsch, W., Häder, D.-P.: Light Perception and Sensory Transduction in Photosynthetic Prokaryotes. Vol. 41, pp. 111–139.

Odom, J. D.: Selenium Biochemistry. Chemical and Physical Studies. Vol. 54, pp. 1–26.

Oelkrug, D.: Absorption Spectra and Ligand Field Parameters of Tetragonal 3*d*-Transition Metal Fluorides. Vol. 9, pp. 1–26.

Oosterhuis, W. T.: The Electronic State of Iron in Some Natural Iron Compounds: Determination by Mössbauer and ESR Spectroscopy. Vol. 20, pp. 59–99.

Orchin, M., Bollinger, D. M.: Hydrogen-Deuterium Exchange in Aromatic Compounds. Vol. 23, pp. 167–193.

Peacock, R. D.: The Intensities of Lanthanide $f \longleftrightarrow f$ Transitions. Vol. 22, pp. 83–122.

Penneman, R. A., Ryan, R. R., Rosenzweig, A.: Structural Systematics in Actinide Fluoride Complexes. Vol. 13, pp. 1–52.

Powell, R. C., Blasse, G.: Energy Transfer in Concentrated Systems. Vol. 42, pp. 43–96.

Que, Jr., L.: Non-Heme Iron Dioxygenases. Structure and Mechanism. Vol. 40, pp. 39–72.

Ramakrishna, V. V., Patil, S. K.: Synergic Extraction of Actinides. Vol. 56, pp. 35–90.

Raymond, K. N., Smith, W. L.: Actinide-Specific Sequestering Agents and Decontamination Applications. Vol. 43, pp. 159–186.

Reinen, D.: Ligand-Field Spectroscopy and Chemical Bonding in Cr^{3+}-Containing Oxidic Solids. Vol. 6, pp. 30–51.

Reinen, D.: Kationenverteilung zweiwertiger $3d^n$-Ionen in oxidischen Spinell-, Granat- und anderen Strukturen. Vol. 7, pp. 114–154.

Reinen, D., Friebel, C.: Local and Cooperative Jahn-Teller Interactions in Model Structures. Spectroscopic and Structural Evidence. Vol. 37, pp. 1–60.

Reisfeld, R.: Spectra and Energy Transfer of Rare Earths in Inorganic Glasses. Vol. 13, pp. 53–98.

Reisfeld, R.: Radiative and Non-Radiative Transitions of Rare Earth Ions in Glasses. Vol. 22, pp. 123–175.

Reisfeld, R.: Excited States and Energy Transfer from Donor Cations to Rare Earths in the Condensed Phase. Vol. 30, pp. 65–97.

Reisfeld, R., Jørgensen, C. K.: Luminescent Solar Concentrators for Energy Conversion. Vol. 49, pp. 1–36.

Russo, V. E. A., Galland, P.: Sensory Physiology of *Phycomyces Blakesleeanus.* Vol. 41, pp. 71–110.

Rüdiger, W.: Phytochrome, a Light Receptor of Plant Photomorphogenesis. Vol. 40, pp. 101–140.

Ryan, R. R., Kubas, G. J., Moody, D. C., Eller, P. G.: Structure and Bonding of Transition Metal-Sulfur Dioxide Complexes. Vol. 46, pp. 47–100.

Sadler, P. J.: The Biological Chemistry of Gold: A Metallo-Drug and Heavy-Atom Label with Variable Valency. Vol. 29, pp. 171–214.

Schäffer, C. E.: A Perturbation Representation of Weak Covalent Bonding. Vol. 5, pp. 68–95.

Schäffer, C. E.: Two Symmetry Parameterizations of the Angular-Overlap Model of the Ligand-Field. Relation to the Crystal-Field Model. Vol. 14, pp. 69–110.

Scheidt, W. R., Lee, Y. J.: Recent Advances in the Stereochemistry of Metallotetrapyrroles. Vol. 64, pp. 1–70.

Schmid, G.: Developments in Transition Metal Cluster Chemistry. The Way to Large Clusters. Vol. 62, pp. 51–85.

Schmidt, P. C.: Electronic Structure of Intermetallic B 32 Type Zintl Phases. Vol. 65, pp. 91–133.

Schneider, W.: Kinetics and Mechanism of Metalloporphyrin Formation. Vol. 23, pp. 123–166.

Schubert, K.: The Two-Correlations Model, a Valence Model for Metallic Phases. Vol. 33, pp. 139–177.

Schutte, C. J. H.: The Ab-Initio Calculation of Molecular Vibrational Frequencies and Force Constants. Vol. 9, pp. 213–263.

Schweiger, A.: Electron Nuclear Double Resonance of Transition Metal Complexes with Organic Ligands. Vol. 51, pp. 1–122.

Shamir, J.: Polyhalogen Cations. Vol. 37, pp. 141–210.

Shannon, R. D., Vincent, H.: Relationship between Covalency, Interatomic Distances, and Magnetic Properties in Halides and Chalcogenides. Vol. 19, pp.1–43.

Shriver, D. F.: The Ambident Nature of Cyanide. Vol. 1, pp. 32–58.

Siegel, F. L.: Calcium-Binding Proteins. Vol. 17, pp. 221–268.

Simon, A.: Structure and Bonding with Alkali Metal Suboxides. Vol. 36, pp. 81–127.

Simon, W., Morf, W. E., Meier, P. Ch.: Specificity for Alkali and Alkaline Earth Cations of Synthetic and Natural Organic Complexing Agents in Membranes. Vol. 16, pp. 113–160.

Simonetta, M., Gavezzotti, A.: Extended Hückel Investigation of Reaction Mechanisms. Vol. 27, pp. 1–43.

Sinha, S. P.: Structure and Bonding in Highly Coordinated Lanthanide Complexes. Vol. 25, pp. 67–147.

Sinha, S. P.: A Systematic Correlation of the Properties of the f-Transition Metal Ions. Vol. 30, pp. 1–64.

Schmidt, W.: Physiological Bluelight Reception. Vol. 41, pp. 1–44.

Smith, D. W.: Ligand Field Splittings in Copper(II) Compounds. Vol. 12, pp. 49–112.

Smith, D. W., Williams, R. J. P.: The Spectra of Ferric Haems and Haemoproteins, Vol. 7, pp. 1–45.

Smith, D. W.: Applications of the Angular Overlap Model. Vol. 35, pp. 87–118.

Solomon, E. I., Penfield, K. W., Wilcox, D. E.: Active Sites in Copper Proteins. An Electric Structure Overview. Vol. 53, pp. 1–56.

Somorjai, G. A., Van Hove, M. A.: Adsorbed Monolayers on Solid Surfaces. Vol. 38, pp. 1–140.

Speakman, J. C.: Acid Salts of Carboxylic Acids, Crystals with some "Very Short" Hydrogen Bonds. Vol. 12, pp. 141–199.

Spiro, G., Saltman, P.: Polynuclear Complexes of Iron and their Biological Implications. Vol. 6, pp. 116–156.

Strohmeier, W.: Problem and Modell der homogenen Katalyse. Vol. 5, pp. 96–117.

Sugiura, Y., Nomoto, K.: Phytosiderophores – Structures and Properties of Mugineic Acids and Their Metal Complexes. Vol. 58, pp. 107–135.

Tam, S.-C., Williams, R. J. P.: Electrostatics and Biological Systems. Vol. 63, pp. 103–151.

Teller, R., Bau, R. G.: Crystallographic Studies of Transition Metal Hydride Complexes. Vol. 44, pp. 1–82.

Thompson, D. W.: Structure and Bonding in Inorganic Derivates of β-Diketones. Vol. 9, pp. 27–47.

Thomson, A. J., Williams, R. J. P., Reslova, S.: The Chemistry of Complexes Related to *cis*-$Pt(NH_3)_2Cl_2$. An Anti-Tumor Drug. Vol. 11, pp. 1–46.

Tofield, B. C.: The Study of Covalency by Magnetic Neutron Scattering. Vol. 21, pp. 1–87.

Trautwein, A.: Mössbauer-Spectroscopy on Heme Proteins. Vol. 20, pp. 101–167.

Tressaud, A., Dance, J.-M.: Relationships Between Structure and Low-Dimensional Magnetism in Fluorides. Vol. 52, pp. 87–146.

Tributsch, H.: Photoelectrochemical Energy Conversion Involving Transition Metal d-States and Intercalation of Layer Compounds. Vol. 49, pp. 127–175.

Truter, M. R.: Structures of Organic Complexes with Alkali Metal Ions. Vol. 16, pp. 71–111.

Umezawa, H., Takita, T.: The Bleomycins: Antitumor Copper-Binding Antibiotics. Vol. 40, pp. 73–99.

Vahrenkamp, H.: Recent Results in the Chemistry of Transition Metal Clusters with Organic Ligands. Vol. 32, pp. 1–56.

Valach, F., Koreň, B., Sivý, P., Melník, M.: Crystal Structure Non-Rigidity of Central Atoms for Mn(II), Fe(II), Fe(III), Co(II), Co(III), Ni(II), Cu(II) and Zn(II) Complexes. Vol. 55, pp. 101–151.

Wallace, W. E., Sankar, S. G., Rao, V. U. S.: Field Effects in Rare-Earth Intermetallic Compounds. Vol. 33, pp. 1–55.

Warren, K. D.: Ligand Field Theory of Metal Sandwich Complexes. Vol. 27, pp. 45–159.

Warren, K. D.: Ligand Field Theory of f-Orbital Sandwich Complexes. Vol. 33, pp. 97–137.

Warren, K. D.: Calculations of the Jahn-Teller Coupling Costants for d^x Systems in Octahedral Symmetry via the Angular Overlap Model. Vol. 57, pp. 119–145.

Watson, R. E., Perlman, M. L.: X-Ray Photoelectron Spectroscopy. Application to Metals and Alloys. Vol. 24, pp. 83–132.

Weakley, T. J. R.: Some Aspects of the Heteropolymolybdates and Heteropolytungstates. Vol. 18, pp. 131–176.

Wendin, G.: Breakdown of the One-Electron Pictures in Photoelectron Spectra. Vol. 45, pp. 1–130.

Weissbluth, M.: The Physics of Hemoglobin. Vol. 2, pp. 1–125.

Weser, U.: Chemistry and Structure of some Borate Polyol Compounds. Vol. 2, pp. 160–180.

Weser, U.: Reaction of some Transition Metals with Nucleic Acids and their Constituents. Vol. 5, pp. 41–67.

Weser, U.: Structural Aspects and Biochemical Function of Erythrocuprein. Vol. 17, pp. 1–65.

Weser, U.: Redox Reactions of Sulphur-Containing Amino-Acid Residues in Proteins and Metalloproteins, an XPS-Study. Vol. 61, pp. 145–160.

Willemse, J., Cras, J. A., Steggerda, J. J., Keijzers, C. P.: Dithiocarbamates of Transition Group Elements in "Unusual" Oxidation State. Vol. 28, pp. 83–126.

Williams, R. J. P.: The Chemistry of Lanthanide Ions in Solution and in Biological Systems. Vol. 50, pp. 79–119.

Williams, R. J. P., Hale, J. D.: The Classification of Acceptors and Donors in Inorganic Reactions. Vol. 1, pp. 249–281.

Williams, R. J. P., Hale, J. D.: Professor Sir Ronald Nyholm. Vol. 15, pp. 1 and 2.

Wilson, J. A.: A Generalized Configuration-Dependent Band Model for Lanthanide Compounds and Conditions for Interconfiguration Fluctuations. Vol. 32, pp. 57–91.

Winkler, R.: Kinetics and Mechanism of Alkali Ion Complex Formation in Solution. Vol. 10, pp. 1–24.

Wood, J. M., Brown, D. G.: The Chemistry of Vitamin B_{12}-Enzymes. Vol. 11, pp. 47–105.

Woolley, R. G.: Natural Optical Activity and the Molecular Hypothesis. Vol. 52, pp. 1–35.

Wüthrich, K.: Structural Studies of Hemes and Hemoproteins by Nuclear Magnetic Resonance Spectroscopy. Vol. 8, pp. 53–121.

Xavier, A. V., Moura, J. J. G., Moura, I.: Novel Structures in Iron-Sulfur Proteins. Vol. 43, pp. 187–213.

Zumft, W. G.: The Molecular Basis of Biological Dinitrogen Fixation. Vol. 29, pp. 1–65.